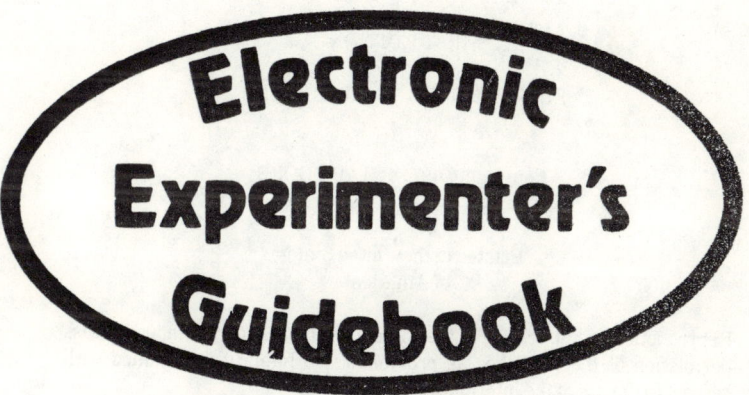

By Don Tuite

Blue Ridge Summit, Pa. 17214

FIRST EDITION

FIRST PRINTING—OCTOBER 1974

Copyright © 1974 by TAB BOOKS

Printed in the United States
of America

Reproduction or publication of the content in any manner, without express permission of the publisher, is prohibited. No liability is assumed with respect to the use of the information herein.

Hardbound Edition: International Standard Book No. 0-8306-4540-3

Paperbound Edition: International Standard Book No. 0-8306-3540-8

Library of Congress ard Number: 74-14322

The "Amphenaut" on the cover is the creation of John Gove, product manager at Bunker Ramo's Amphenol Industrial Div., Chicago, Illinois.

PREFACE

This book is intended as a kind of bridge. On one side, there's you. You know a little about electronics, but want to know a great deal more. You've read through some books on basic theory; so you know about Ohm's law, and you know what the lines and squiggles on a schematic diagram mean, even if you aren't always sure why all of them are there. You may have enjoyed putting together a hi-fi or test equipment kit.

On the other side of the bridge are all the home-built projects and roll-your-own pieces of gear that appear in hobby magazines and books. Beyond that lies the wide world of electronic projects that you can pull together out of ideas in your own head and create from scratch with your own hands.

Somehow, there's a gulf between the place where you stand now and all those fascinating homebrew projects. Unpacking the little kit parts from the box with the Benton Harbor postmark is different from facing the clerk at the radio parts store with a handwritten list of parts and no idea of what to say when he tells you that a certain item is indefinitely out of stock! Somehow, too, you have the feeling that if you unwittingly make a mistake, you will wind up with $30 worth of junk parts. "See these cufflinks?" you envision yourself telling your friends, "They used to be $5 transistors!"

What you need is a bridge—a bridge to let you cross from where you are now to a place where you feel confident to undertake a project from a magazine or book, and maybe, ultimately, projects entirely of your own creation. This book is intended to be your bridge.

When you have finished this book, you should know quite a bit about the tools required to make a neat, functional project; the various ways of finishing projects off to give them that "professional" appearance; scientific methods of troubleshooting gear; and finally, you will understand some of the considerations involved in making substitutions when the exact component specified isn't available. To help make the text clearer, there are eight original projects that illustrate the concepts in each chapter. You may want to build some or all of them, or you may just want to read through them and look at the photographs to see how the ideas are embodied in real hardware.

Building your own electronic gear is a rewarding pastime, and this book is intended to be a pleasant introduction to that pastime. So relax, take your time, and enjoy yourself.

Don Tuite

CONTENTS

1 *Foundation For Electronic Construction* — 9
Recommended Tools—Soldering—Layout—Reminders

2 *Construction Methods* — 27
Perforated Board With Push-In Terminals—Perforated Boards With Adhesive-Backed Circuits—True Printed Circuits—Metal Chassis—Generally Applicable Techniques—Multi-Strobe Project—12V DC to 110V AC Inverter Project—Triac Controller Project

3 *Finishing Touches* — 70
Cases and Enclosures—Laying Out a Control Panel—Legible Labels—Selecting Knobs, Switches, and Indicators—Making Your Own Meter Scales—Randomizer Project—FET-Set Shortwave Receiver Project

4 *Troubleshooting Your Projects* — 104
Troubleshooting Equipment—Systematic Troubleshooting—After You Find The Trouble—Capaci-Bridge Project—BJT-FET Transistor Checker Project—Theory of Transistor-Checker—Signal Generator Project

5 *Making Successful Substitutions* — 149
Resistors—Capacitors—Coils and Transformers—Diodes—Bipolar Transistors—Field-Effect Transistors—Unijunction Transistors—Integrated Circuits

Appendix A Electronic Color Coding — 165

Appendix B Electronic Symbols — 174

Index — 178

Foundation For Electronic Construction

The best place to begin is with the tools you will need for building your projects. The tools listed below are just about all you will need to tackle any of the projects in this book, and in fact, just about any project you will come across anyplace else.

RECOMMENDED TOOLS

You won't need all of these recommended tools for every project in this book, but if you will be doing much electronics experimenting, you will need each of the tools on the list. There are some other tools like socket punches, tin snips, or heavy-duty soldering irons that you might find a need for from time to time, but you're not likely to need them right off, so it's best to wait until you find yourself with a specific project that calls for them before buying them.

Pliers

You'll need **long-nose** pliers for getting into tight spaces, fastening wires to solder lugs, and retrieving dropped parts from crowded corners of chassis. It's handy having two sizes, but if you can only afford one, get the longer variety. Before you buy a pair of long-nose pliers, hold them up to the light and look through the closed jaws. The less daylight you can see between the jaws, the better. Don't reject a pair that has a ridged gripping surface at the tip and daylight showing the rest of the way down to the hinge, however (Fig. 1-1). If the jaws meet evenly all over the ridged area, you've found a good pair of pliers.

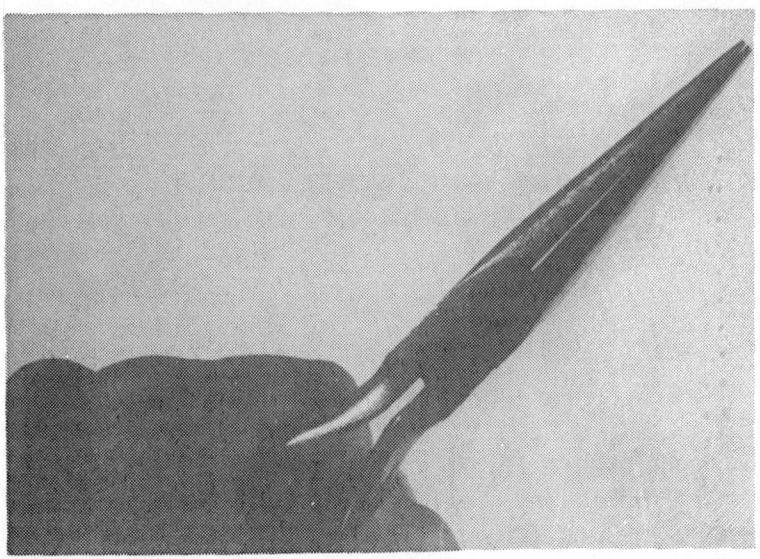

Fig. 1-1. Held up to the light, pliers should show little if any gap between jaws. This pair has a ridged gripping surface at the tip, so the gap that appears below that surface is acceptable.

No doubt your long-nose pliers will have a cutting area down near the base. This won't do for all your needs, though. You may want to clip a wire down in a cramped space someday. For this you need **diagonal side-cutting pliers,** or **"dikes."** When you are working on your projects, do yourself a favor and don't try to cut oversize wire with an undersize pair of dikes. "Oversize" wire is anything over No. 12 (AWG) for most dikes. When you want to cut oversize wire, first nick it with a knife where you want to cut, then bend the wire back and forth a few times at that place. The wire will break cleanly right where you want it to, and you'll have avoided dulling your wire cutters.

After you've cut a length of wire, you still have to strip the insulation off. The best tool for this is a nightmarish-looking pair of pliers with a vice-like arrangement in one set of jaws to hold the body of the wire and a set of knife edges in the other jaw with notches in them to fit different sizes of wire. These tools never nick wires and they're fascinating to watch, but they're priced anywhere from $7 to over $20.

One of the more desirable kinds of **wire strippers** is a very simple type with notched shear-type blades. There's also a combination tool for wire stripping and crimping solderless terminals. The first type is very handy to use, is readily adjustable to different wire sizes, and is cheap. The second type is more expensive, but there are some household uses for the solderless connectors, so it isn't a bad investment. Whatever you do, don't rely on a pocket knife for your wire stripping needs. You'll waste a lot of time if you do, and sooner or later, you will cut a piece of wire too short, with no replacement in your junk box and just a half-hour too late to get to the store to buy a replacement.

As a general rule, you should never use a pair of pliers on a nut. Use a wrench, preferably one of the right size, or at least a good adjustable wrench. Like all general rules, however, there are times when this one is well broken. Sometimes it is very handy to have a pair of 10 in. **Vice-Grip** pliers. These pliers will also prove themselves invaluable for holding small work while you are soldering it, and for hundreds of other uses. They are a good investment.

You will need at least two **flat-bladed** screwdrivers, one with a ¼ in. blade for most ordinary screws, and one with a ⅛ in. blade for the setscrews in most knobs and for tight work. A set of **jeweler's** screwdrivers may prove handy, but it is not an absolute necessity. You will need at least one **Phillips head** screwdriver.

Some knobs do not have slotted-head setscrews, but have Allen head setscrews instead. There are several sizes of these, and it is best to simply buy a set of **Allen** (or **hex key**) wrenches.

Wrenches

The two sizes of hex-head nuts you will encounter most frequently are the ¼ and ½ in. sizes. The ¼ in. size is found most frequently on nuts for 4-40, 6-32, and 8-32 screws. It is also found on a variety of sheet metal screws. The ½ in. hex nut is used for the bushings of most potentiometers and toggle switches. It is handy to have a nutdriver for ¼ in. hex nuts and an open-end wrench for ½ in. hex nuts. Again, do not mangle these nuts with some kind of pliers—use a **wrench** to hold them. Nothing is more certain to spoil your attitude about a project than slipping while trying to tighten a nut with the pliers and scratching the finish on a neatly painted control panel.

From time to time, you will encounter other sizes of hex-head nuts. For these occasions, you will need either a small, adjustable open-end wrench or an assortment of nutdrivers and wrenches in various sizes.

Electric Drill and Bits

You can no more make a project without drilling holes than an omelet without eggs. With current prices on hand and electric drills, it makes sense to buy an **electric drill**, preferably one with a variable-speed motor. There are some important advantages to the lower speed ranges available with modern electric drills. With slow-speed capability, you can start a hole right where you want it, without center-punching it first, and you know that the bit isn't going to wander all over the chassis. Also, brittle materials like plastic, Bakelite, and printed circuit board drill better at low speed, with less chipping and cracking.

When you buy **bits**, you'll probably find it doesn't cost much more to buy an assortment of 10 than it does to buy one each of the sizes you really need. However, it is worth knowing that you can get by with six basic sizes and some tiny bits for printed circuit boards. The six sizes you will absolutely need are the three sizes needed to pass the most commonly used screws, plus ¼, ⅜, and ½ in. sizes.

There are two ways of identifying bits: by their number sizes, which are related to their dimensions in thousandths of an inch, or by their diameters, expressed in fractions of an inch. The most common fractions-of-an-inch types are graduated in multiples of 1/64 of an inch and do not correspond to any of the numbered sizes. Therefore, for each screw size, there are two possible sizes of drills to buy, depending on how the drills are designated. Thus, for clearing 8-32 screws, drill a hole using either a No. 18 or a 3/16 in. bit. For a 6-32 screw, use a No. 28 or a 9/64 in. bit. For a 4-40 screw, use a No. 33 or a ⅛ in. bit.

You'll need a ¼ in. bit to make holes for passing the shafts of the most common sizes of variable capacitors. The ⅜ in. bit will be needed to make holes for the bushings of potentiometers and toggle switches. It also makes the minimum size hole required to pass the jaws or the nibbling tool that will be discussed in the next section. The ½ in. bit will make holes that will pass pilot light assemblies and some kinds of binding posts. Both ⅜ and ½ in. drills can be bought with ¼ in. shanks for use in the most common electric drills. When you use these oversize bits, however, bear in mind that you are calling on your drill to do more than it was designed to do. Expect some chatter, and brace the work firmly. Keep your fingers out of harm's way. Don't overheat your drill with heavy stock.

Most radio stores and supply houses carry special bits for printed circuit work. These bits are exactly the right size to pass the different gages of wire you will be using. They are also designed with a different cutting angle than metalworking drills. This allows them to drill brittle printed circuit boards more easily, and reduces the chance of these delicate bits breaking.

Nibbling Tool

If ever there was a tool to gladden the heart of an electronics experimenter, the nibbler (Fig. 1-2) is it. The nibbler works on the basis of a small, tool-steel (very hard) shears and a handle with a tremendous mechanical advantage. In operation, the head of the tool is passed through a ⅜ in. hole and the jaws are positioned for the first cut. From there the tool literally nibbles any desired shape in metal up to 18 gage mild steel or 16 gage aluminum.

Saws

Not every piece of metal or plastic can be nibbled. The shafts on variable capacitors and potentiometers, for instance, must be trimmed to size with a hacksaw. Often, plastic boxes for projects are too thick to fit in the jaws of the nibbler. This is where a keyhole saw comes in handy. One of the best and least expensive keyhole saws consists of an X-acto knife handle with a specially designed keyhole saw blade (Fig. 1-3). It takes a little work to find these blades sometimes. The surest place to find them is in a store that sells to model airplane hobbyists.

Fig. 1-2. The nibbling tool is a godsend for cutting odd-sized holes in panels and chassis. The panel in the picture is covered with masking tape to protect the finish and provide a surface for marking positions of holes to be drilled.

Knives

It's nice to have a pocket knife that you can carry around and have handy for odd repairs, but for workshop use, the **X-acto** knife with X-acto's No. 11 blade is the handiest thing you will find. The reason for this is the extreme sharpness of the blade. This makes it easy to cut through insulation with very slight pressure, which results in fewer nicked and broken wires. One caution: The blades of these knives do wear out fairly quickly; as soon as it seems to be happening to the blade you're using, throw it out and replace it with a new one. This will save you considerable frustration, particularly when you are trimming patterns for etched circuits.

Files

Drilling holes, or sawing, or nibbling—whatever metal-cutting technique you're using—you are bound to leave burrs and rough edges. Therefore, it is absolutely necessary to file away burrs and smooth edges before painting. You will need large and small sizes of **round, half-round,** and **flat** files, as well as a **rat tail** file. Files do not cut well when they are clogged with metal, so you will also need a wire-bristled brush to clean your files. A ½ in. **hand reamer** is a handy tool to have also, for enlarging holes to odd sizes, since it will do the job faster and leave a rounder hole than a file.

A final essential for electronics is a **soldering gun** or **iron**. Proper soldering is so important to successful completion of a project that soldering tools are treated separately in the following section.

SOLDERING

Proper soldering is a very easy art to learn. However, more kits and projects are ruined by bad soldering than any other cause.

Tools

There are at least two types of soldering guns and two types of soldering irons. It is difficult to decide when to call a tool an **iron** and when to call it a **gun**, but for our purposes, an iron is any soldering implement that is on as long as it is plugged in, and a gun is any soldering implement with a trigger-type on-off switch.

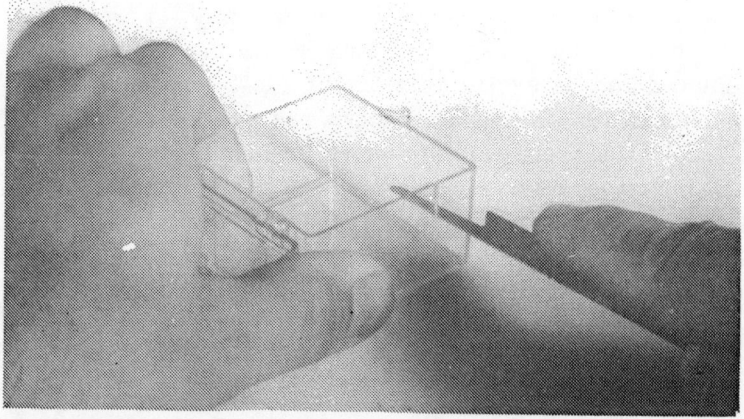

Fig. 1-3. Keyhole saw is essential for making odd-shaped holes in plastic. Blade pictured is available from model airplane hobby shops.

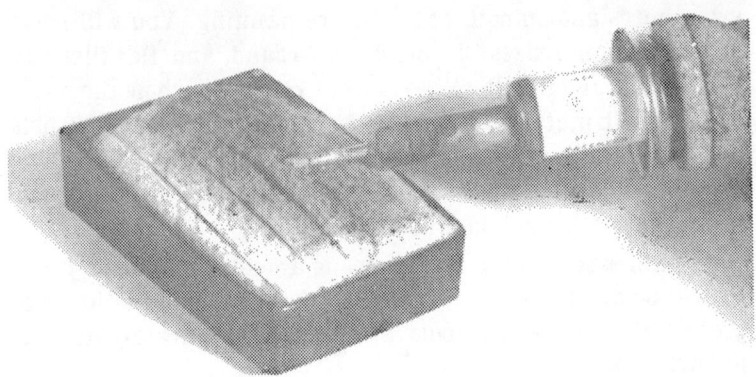

Fig. 1-4. Wipe the tip of your soldering iron frequently to keep it shiny and free of scale. The sponge in the photo is made specifically for this purpose.

The most common kind of gun has a tip that is essentially a loop of wire. This is connected to the secondary of a step-down transformer inside the gun. Another type has an enclosed heating element with a resistance that increases with temperature. This is a more efficient gun, since the current it draws varies as the heat load on the tip. Its disadvantage is that replacement tips are more expensive than tips for the loop type. In fact, an emergency tip can be made for this gun out of any stray piece of heavy wire that may be in the junk box.

The advantage of the gun is that it can be left safely plugged in all the time without burning out the tip or creating a burn hazard. The gun's disadvantages are the warmup time required and its excess heat output. Most guns put out too much heat for delicate transistor work.

Soldering irons can be divided into low-wattage **pencils** and high-wattage **irons**. For anything except soldering to a heavy chassis, the pencil is the only iron suited to electronics work. A soldering pencil in the 45W range is an almost ideal soldering tool, but for very fine work a 37½W pencil is even safer.

Preparing Soldering Tools

To make a good solder joint, the tip of the soldering tool must be properly prepared. You cannot get a good solder joint using a tool with an eighth of an inch of scale glowing cherry red on its tip. A lot of kits and projects go wrong right at this point. If you are starting with a brand new tip on your soldering tool, read on. If the tip of your soldering tool is crusty and corroded, file it down to bare copper, or replace it, and then read on.

Start with a brightly tipped soldering tool and heat it up to its operating temperature. Then take your solder and liberally coat the entire surface of the working end of the gun or iron. From now on, whenever you use the soldering tool, wipe it frequently with a rag or moist sponge to keep the solder on the tip bright and shiny (Fig. 1-4). When you are through with your gun or iron, wipe the tip as it is cooling down to keep the tinning of solder on it shiny. This effort will repay you many times over in fast heating and good solder joints. If you have some money to spare, there is a chemically treated sponge in its own plastic holder that can be bought in most radio specialty shops and from the mail order houses. The sponge has a number of slits in its top which are handy for wiping the tip of your gun or iron, and the plastic holder has a nonskid base that allows easy one-hand operation.

Solder

First, **never use acid core solder.** The only kind of solder to use in electronics work is rosin core. Solder is an alloy of tin and lead. The concentrations of tin and lead in the alloy determine its characteristics. The ideal alloy is 63-37 (63 percent tin, 37 percent lead), or **eutectic** solder. This has the lowest possible melting point of any tin-lead solder, and it is nearly impossible to make a "cold" solder joint with this material. Eutectic solder isn't very common, however. The most frequently encountered alloy is 60-40. This has a satisfactory melting point and resistance to cold solder joints. Some 50-50 alloy solder is available, but its use should be avoided.

Generally, solid solder should not be used for radio work. Solder with a rosin core (or cores) carries just enough rosin with it to vapor-clean each joint as it is made. It is possible to apply rosin flux separately and use solid solder, but the technique is messy. There is one case in which solid solder and a separate paste flux is used—when it is necessary to solder to aluminum. For most low-frequency work, chassis grounds are achieved with solder lugs and screws. For high-power or high-frequency work, however, it may be necessary to solder directly to an aluminum chassis. With ordinary rosin core solder this is impossible, because the oxide coating on the aluminum resists the cleaning effects of the rosin flux. However, there is an aluminum-soldering flux that can be purchased that will clean away aluminum oxide and allow solder to flow. For this one purpose, solid (coreless) solder should be used.

Preparation for Soldering

The first thing to consider in preparing a joint for soldering is what you are soldering to. The basis for any good solder joint is some firmly fixed terminal—either a tube or transistor base pin, a solder lug, a tie point on a terminal strip, a push-in terminal on a piece of perforated board, or a pad on a printed circuit. With few exceptions, one never connects two components except by soldering them to a fixed support.

If one of the components to be soldered is an insulated wire, it will be necessary to strip the end of the wire. This is best done with a stripping tool of some kind, although in a pinch a knife will do. Especially with solid wires, it is very important not to nick the wire. Nicking a wire creates a stress concentration at the nick. You can demonstrate to yourself how serious this is by taking an ordinary copper wire and making a small nick in it. You will find that only four or five sharp bends will cause the wire to break at the nick.

Once you have the ends of your wires and components prepared, the next step is to fasten them firmly to the support point (Fig. 1-5). Bend the wires over and crimp them slightly. The mechanical strength of the joint must come from the method of attachment, not from the solder. The purpose of the solder is only to assure a complete electrical connection, nothing more! On printed circuits, push the lead through the hole and bend the wire over against the copper side of the board to secure it.

Next, **before applying any heat,** ask yourself, what will the heat do to the components? If any of the leads you are soldering come from a transistor or diode, the heat of soldering could damage the part. To protect the part, hang an alligator clip or paper clip from the wire lead, or lean a screwdriver blade or the jaws of your pliers against it to provide a heatsink between the joint to be soldered and the part itself (Fig. 1-6).

Now you are ready to apply heat to the joint. The object is to heat the joint up so that when solder is touched to it, it will flow (Fig. 1-7). Note that you **do not** apply solder to the iron and let it flow down onto the joint. Sometimes, it **is** convenient to apply a little surplus drop of solder to the tip of the tool to aid in heat flow, but the bulk of the solder used should be applied directly to the joint.

You can tell that the solder is flowing properly by observing the way it covers the wires and support. It should wet

Fig. 1-5. Wires must be firmly fastened to the support points before solder is applied.

19

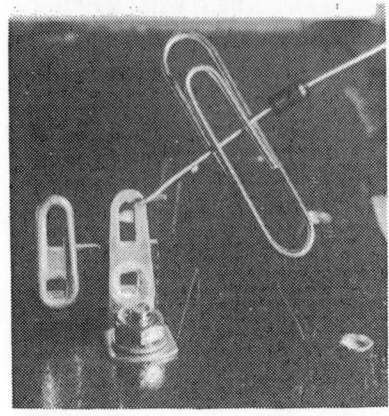

Fig. 1-6. Always use a heatsink when you solder to the leads of a semiconductor device. Here, a paper clip serves as an improvised heatsink.

all the surfaces it comes into contact with, exactly as if it were a drop of water. If it stands up and beads on the surface, then the surface isn't yet hot enough and the solder connection is incomplete. This is the most frequent cause of the **rosin joint** type of failure. At first glance, the joint looks solid, yet all that is holding the parts together is a film of rosin flux. Figure 1-8 shows a typical rosin joint.

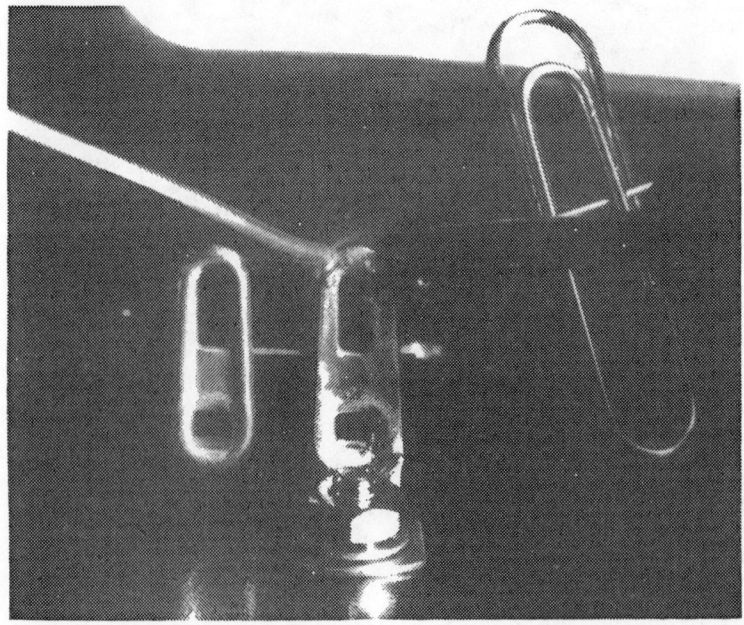

Fig. 1-7. Heat the joint and apply solder to it; do not apply solder to the tip of the iron directly.

Once the solder has flowed onto the joint, remove the heat; do not move any wires in the joint. If you allow any of the wires to move, you will end up with the dull gray surface on your solder that is characteristic of a cold solder joint. Cold solder joints can plague you with high resistance in your connections, or even worse, they can work as diodes to cause unwanted audio from nearby radio stations to appear in audio amplifier circuits. Figure 1-9 shows a cold solder joint.

Do not use too much solder. Use just enough to wet the whole joint. Globs of solder can interfere with pins in sockets and can cause unintended shorts. Figure 1-10 shows a proper solder joint.

Sometimes it is necessary to solder wires inside pins for different kinds of connectors. There is a technique for this that can save a lot of frustration. The technique is based on the **surface tension** of the solder and the principle of **capillarity**.

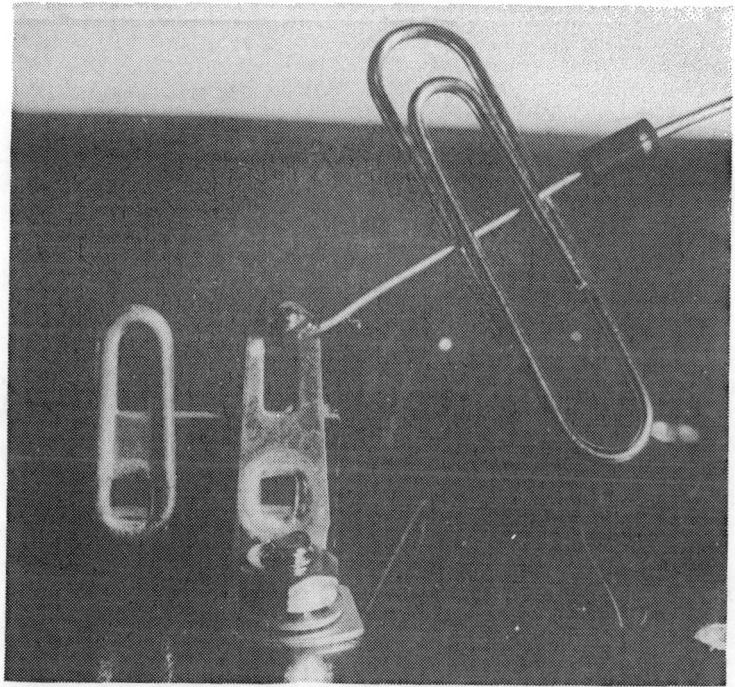

Fig. 1-8. A rosin joint—the connection was not hot enough when solder was applied. The only thing holding the parts together is a layer of rosin flux, while the solder sits uselessly on top. This high-resistance connection can be very hard to spot.

21

Fig. 1-9. A "cold" solder joint. The wire was moved while the joint was cooling, and the lead and tin in the solder did not alloy but crystallized out separately. This emphasizes the importance of making a firm mechanical connection before applying solder.

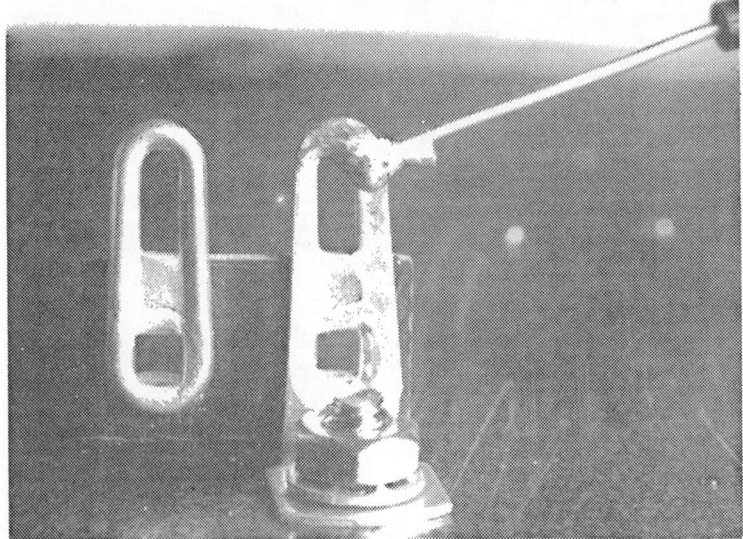

Fig. 1-10. A good solder joint is characterized by a smooth, shiny surface and a minimal amount of solder. Big globs of solder can run down and cause shorts. Note that the solder does not bead up, as it does in a rosin joint, but appears to wet the surfaces it binds.

Fig. 1-11. When soldering to pins, point the opening downward and apply heat to the side of the pin. Molten solder will be sucked up into the pin by capillary action. When the solder has cooled, cut off the excess wire.

The way to solder these pins is **upside down.** If you try to solder them right side up, you will run solder down the outside of the pins and make it difficult to insert them into the holes in the socket. The trick is to use a nearly dry soldering iron and heat the pin until it is hot enough to melt solder. Then apply your solder to the hole at the tip of the pin and it will be sucked up and inside by capillary action (Fig. 1-11). Meanwhile, no solder will adhere to the outside of the pin.

It helps to be able to clear solder out of pins on plugs and holes in terminal lugs when you have to. The process is simple. Just hold the part in a pair of pliers and heat it until the solder is molten. Then quickly strike the part or the base of the pliers on the table and the solder will shoot out (Fig. 1-12). Be careful it doesn't splash on **you**, though. A variation of this technique is to blow sharply on the molten solder. This works better with solder lugs than with pins, however, since the surface tension in the pins is quite high.

LAYOUT

It is always best to follow the same arrangement of leads and components described in your kit manual or in the article

Fig. 1-12. To remove solder from pins, hold plug with pliers and apply heat to pin until solder is molten. Then strike side of pliers sharply on table top and liquid solder will shoot out. The beads of hot solder can bounce wildly, so protect your eyes!

describing your project. In some cases—for instance, in many circuits in semiconductor handbooks and manufacturers' application notes—there is no information given as to the physical layout of the project. In these cases, there are a few simple rules you should keep in mind:

1. Be logical. Use the schematic diagram as a starting point. Position the components between the ground bus and V+ line. The closer your layout is in appearance to the

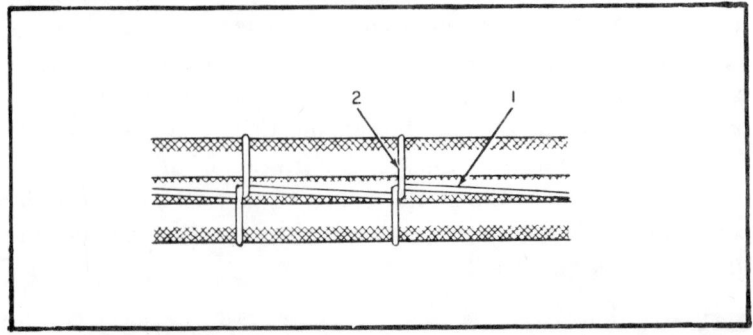

Fig. 1-13. Use waxed cord or heavy thread and make ties as shown to bundle individual wires into cables. Note that if the cord should break at point 1, the loop (2) probably would not loosen.

symbols on the circuit diagram, the fewer mistakes you will make and the easier it will be to troubleshoot your project.

2. If you are using 60 Hz house current for power, keep all lines carrying the current well away from signal leads. If you must make a long run of signal lead, use shielded wire, or for shorter runs, twist the signal lead and its associated ground lead together. Power supply leads in high-power ham transmitters should be shielded.

3. Avoid running leads carrying two different frequencies (rf and i-f, or rf and audio) closely parallel to each other.

4. Be careful about **lead dress** (positioning). Wires carrying radio frequencies should be as short as possible, with no sharp bends. All other wires should be bundled into cables (Fig. 1-13) as much as possible and routed along edges of circuit boards or chassis, or parallel to edges. For longer runs, use tiedown anchors.

5. Group controls for convenient operation. For critical settings such as main tuning controls or speed controls, larger knobs are easier to handle.

6. Do not let components **hang** by their leads. This is an invitation to vibration or shock failure after a few hundred hours of operation.

REMINDERS

Take the time to use heatsinks. It is surprising how little heat in the wrong place is required to destroy a semiconductor device.

In addition to using heatsinks, be careful with hot soldering irons. They may radiate enough heat to destroy a nearby transistor. Be careful also not to melt the insulation on wires near where you are soldering. At the very least, it makes the final product look unsightly. At worst, you may have a very hard-to-locate short circuit.

Avoid dropping any components. Hard as it may be to believe, a drop of a few feet onto a hard surface can subject a device to a shock pulse of thousands of times the force of gravity. This can have fatal effects on transistors and diodes.

Be especially careful when handling MOSFETs that do not have integral gate protection. Static charges that build up on your body from contact with clothes can punch through the oxide insulating the gates of these devices from their channels. Always use transistor sockets with unprotected MOSFETs. Just before picking up an unprotected MOSFET, discharge yourself to a large metal object.

If all the preceding **dos** and **don'ts** haven't intimidated you, you are ready to take on the different methods of construction—perforated board and terminal, printed circuit, and metal chassis with point-to-point wiring.

Construction Methods

There are four basic electronics construction methods that we will discuss in this chapter. Two of the methods, metal chassis and perforated board with push-in terminals, use point-to-point wiring. The other methods use circuits on the surface of insulating boards. There are advantages and disadvantages to all the methods. Your selection of a particular method for a specific project will depend on the nature of the project and the resources you have available.

The different methods of construction are demonstrated in the projects in this chapter.

PERFORATED BOARD WITH PUSH-IN TERMINALS

In the very early days of radio, circuits were assembled on top of wooden bases. Brackets were used to hold controls and tube sockets above the surface of the wooden base, and wires and components were run from terminal to terminal. The similarity of these wooden bases to kitchen breadboards (no doubt, many actual breadboards were used by early home experimenters) led to the name of method of construction. The term "breadboard" is still used to indicate a construction technique in which a prototype circuit can be quickly assembled and easily modified.

The perforated boards with push-in terminals that we have today (Fig. 2-1) are descendents of the early experimenters' wooden breadboards. There are two hole sizes, 0.093 and 0.062 in., and several different kinds of hole spacing. There are different kinds of push-in terminals that are used with these holes. Most of the time, you will be using the kind

Fig. 2-1. Perforated board with push-in solder terminal. Other types of push-in terminals are designed for use without solder to permit quick connection and disconnection of experimental circuits.

that is designed to grip a wire while you solder to it. You may also, however, find use for the quick-disconnect type for patching together prototype circuits before you decide on a final design.

The advantages of this kind of construction are the short amount of time required to go from idea to finished circuit, the ability of the boards to be reused almost indefinitely, and the ease with which modifications or corrections can be made. The principal disadvantages are the lack of heat dissipation compared to metal chassis construction and the relative bulk of the finished project compared to printed circuit construction.

The technique for building circuits with perforated board and push-in terminals is simple and straightforward.

First, get an idea of how much space the circuit will take up by laying out your components on the board in the pattern in which you will be assembling them.

Then, sketch the layout to scale. An easy way to go about this is to first lay a piece of paper down on a plain piece of perforated board and then to rub lightly over the surface of the paper with a soft pencil. This will leave you with a pattern matching the holes in the board, which you can then draw your layout upon.

Locate the holes you will need for control shafts and standoffs. Drill all of these holes before you insert any terminals.

Using your layout sketch, insert the push-in terminals into holes as required. Do not try to connect too many wires to a single terminal. Use two terminals in adjacent holes connected by a short wire.

Now mount your controls and standoffs and all the other hardware such as transformers and sockets that you will need. Be sure to use lockwashers under all nuts and on all control shaft bushings.

You are now ready to install your components and solder their terminals. Solder as you go. As soon as you have all of the wires for a particular terminal in place, solder them. This will help to avoid the frustration of wires springing loose.

If you are doing some original experimenting, there is a technique that can be used with perforated board that is very handy. Use a relatively large piece of board and run buses down its length for each important node in your circuit. You would have one bus at the top of the board for V+; another along the bottom for ground; another, if appropriate, for an agc line; and so forth. Label each bus so you will be able to identify it easily. With this kind of arrangement, it becomes a simple matter to swap components and configurations to determine the best circuit.

PERFORATED BOARDS WITH ADHESIVE-BACKED CIRCUITS

One very handy construction technique uses conductors with pressure sensitive adhesive backing, marketed under the trade name "Circuit Stik." There are a variety of pads with proper spacing to allow immediate insertion of transistors in standard cases, integrated circuits in flatpacks and cans, and discrete components. There are several sizes of copper ribbon with adhesive backing that are used to connect the pads in the finished circuit. The final product is a board that looks very much like a true etched or printed circuit board (Fig. 2-2).

The advantages of this construction method are neatness, speed of layout, and low lead inductance. You can also make

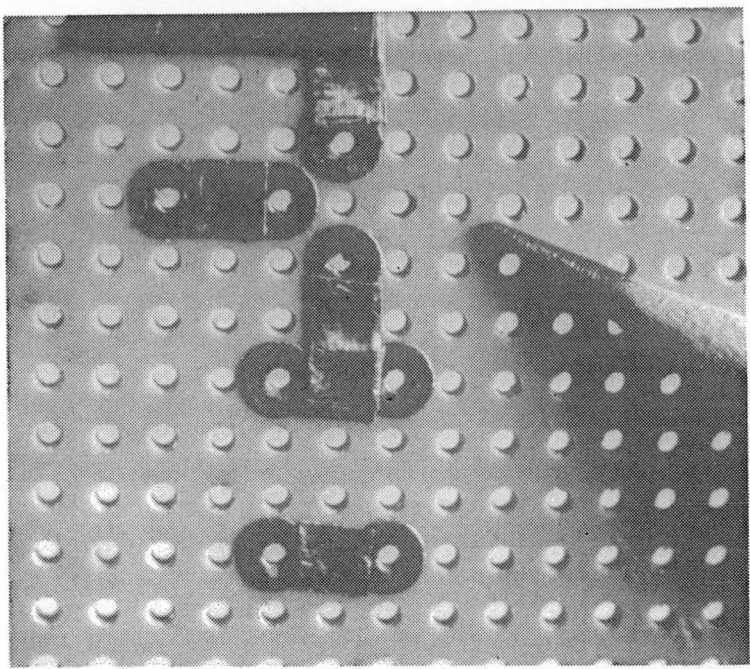

Fig. 2-2. Adhesive-backed copper pads and ribbon can be used to make circuit boards with many of the advantages of true etched circuits.

corrections and modifications to your circuits. Laying out a few projects with these adhesive pads and strips will give you some good practice in laying out printed circuits—without all the fuss and bother associated with acid etching. A final advantage of this method is that you can easily transfer a successful layout to a printed circuit **master** for group projects.

The main disadvantage of these stick-on circuits is that they are relatively expensive. By the time you buy the board, copper strip, universal pads, and transistor pads for even a simple project, you will find that you already have a bill for $10 to $15. Another disadvantage is that sometimes pads and strips do not stick as well as they should and come loose from the board during soldering. But if you are careful to burnish the stick-on elements in place well and not to get oil from your skin on the surface of the board, peeling will not be a serious problem.

The first step in building projects using this technique is to sketch the layout you plan to follow. If you find that you cannot make the circuit without crossing wires, you can run the crossing conductors on the back side of the circuit board. Make your sketch roughly to scale.

Then begin to lay out your pads and connecting strips on the perforated board. Use the real parts for templates to determine the proper spacing between pads. For instance, you should take a resistor of the size you are using and bend its leads to the proper shape and use this to find the exact holes in the perforated board on which to place your pads. Remember that each lead from each resistor, capacitor, etc. requires its own hole.

There are two ways to mount components like resistors and capacitors. You can install them with their bodies flat against the board or, to really save space, you can use a form of what engineers call "cordwood" packaging. In cordwood packaging, one of the two axial leads of the resistor or capacitor comes straight out from the body of the component and passes through one hole in the board. The other lead is bent back along parallel to the body and passes through another hole very close to the first. This practice permits components to be crowded together in a very dense manner (Fig. 2-3). The image of many components stacked side by side is what originally gave the mounting method its name. Another variation of cordwood packaging uses two boards, one above the other, with circuits on each board and components stacked between them. Designing these circuits is very time-consuming, but computer manufacturers, in particular, have found it rewarding.

After you have laid out all of your pads and interconnecting strips, cover the entire board with a piece of

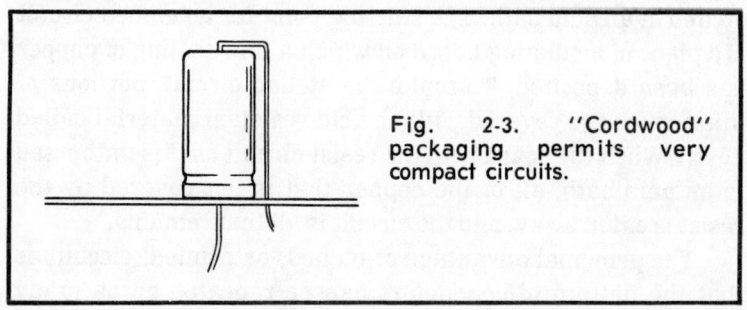
Fig. 2-3. "Cordwood" packaging permits very compact circuits.

paper and burnish all the circuit elements heavily with a burnishing tool or the plastic cap from a ballpoint pen. This burnishing is very important to the sticking of the circuit elements, so make sure you bear down heavily on each one.

Now, as long as you have a piece of paper on top of the board anyway, take a soft lead pencil and very lightly rub over the entire surface of the board. This will cause an image of the circuit to appear on the paper. Remove the paper from the board and transfer your initial sketch to the image of the circuit on the paper. This will tell you whether you have, in fact, provided for all of the leads from all of your components. If you have not, it is a simple matter at this point to go back and add what you need, taking care to burnish each new addition.

Following your new sketch, install your components. Push the leads of each component through the board from the top side until the body of the component is flush with the surface of the board. Then bend the leads of the component on the circuit side over until they are flush with that surface of the board, and cut all but an ⅛ in. or so off with your dikes.

Be careful not to apply any more heat than necessary, and solder each bent-over wire to its pad. Then go back and solder each discontinuity between pad and strip, pad and pad, or strip and strip. This is necessary to make sure that there are no high-resistance joints caused by the adhesive.

TRUE PRINTED CIRCUITS

The term "printed circuit" is actually inaccurate. So-called printed circuits are not made by any process resembling the one used to make this book. Instead, the circuits are **etched** in an acid bath. The starting point for an etched circuit is a piece of insulating board on which a thin coating of copper has been deposited. To make the etched circuit, portions of this copper are covered with an acid-resistant material called **resist**. When the board with the resist circuit on it is immersed in an acid bath, all of the copper that is not covered by the resist is eaten away, and the circuit is all that remains.

The principal advantage of etched, or printed, circuits is that the pattern of conductors can be repeated on as many circuit boards as required. This makes this construction technique ideal for large production runs. For the home hobbyist, this repeatability can also be an advantage, since groups of people can all take on the same project and be assured that they will all achieve similar results.

Other advantages of etched circuits are that they are neat, very reliable, and have low lead inductance. At UHF and microwave frequencies, they can be used in stripline assemblies.

The disadvantages of etched circuits are that they are relatively time-consuming to prepare, the chemicals required are expensive and slightly hazardous, and there are no convenient ways of correcting mistakes on etched circuit boards. If you goof, you have to return to square No. 1 and start over.

There are three common methods used by home experimenters for preparing etched circuit boards; all three have many steps in common. The three ways of making etched circuits are:

- Direct application of resist
- Sensitized board and photo negative
- Sensitized board and mechanical negative

In the first method, resist is simply painted or drawn on the surface of an untreated copper-clad board. In the other two methods, the copper surface of the board is treated with a photosensitive chemical. When these boards are exposed to light through a high-contrast negative and developed, the parts of the board exposed to light are acid-resistant, while the unexposed portions allow the acid to etch the copper.

First Steps

For all three processes, the first thing you must do is to lay out your circuit on a piece of paper. Many magazine projects provide a full-size pattern for experimenters to trace or copy. If you are starting from scratch, you will have to lay out your circuit to scale, taking careful note of the size of all your components and the positions of their leads.

Many builders limit the width of their conductors and use circular pads for soldering component leads. Here are some

Fig. 2-4. Special felt tip pen draws lines in resist ink directly onto copper-clad board. After etching, it is necessary to remove the resist with a solvent before soldering.

guidelines for this technique: Five amps is the most that a single 1/16 in. conductor should be made to carry. No conductor should be narrower than 1/32 in. Minimum spacing between conductors should be 1/32 in. Round pads typically have a 3/32 in. radius. If you are making negative artwork, allow at least ¼ in. on each edge to allow for clamps during processing. Avoid sharp bends; use **curves** instead.

An alternative to narrow conductors and pads is to leave large areas of conducting copper and etch away only enough copper to form the circuit. The way to prepare your artwork if you are using this technique is to first lay out the circuit roughly with lines indicating conductors and then to build up a series of rectangles around these lines to form large areas of conductor.

The choice of which system to use comes down to a tradeoff in complexity. Simpler circuits are more easily prepared with the block method. As the circuits get more complicated, it becomes harder to deal with the blocks and not make mistakes. As you become more experienced, you will be better able to judge which kinds of projects to subject to which layout technique.

Direct Application of Resist

After you have made your full-size layout, check it for errors, and when you are sure it is correct, tape it and a sheet of ordinary carbon paper to a clean, untreated copper-clad board. Trace over the entire design to transfer it to the board. When you remove the pattern and the carbon paper, you should find the pattern transferred to the copper surface of the board.

The next step is to apply resist to the board. You can use liquid resist and a fine-tipped artist's brush, or you can purchase a special felt tip pen made with resist ink (Fig. 2-4). With this pen, you can simply and easily draw your conductors and pads. Besides the commercially available resists, you can use exterior enamel paint or even India ink.

Once you have the pattern on the board, you can follow the steps in the section titled "Etching" (below).

Sensitized Board and Photo Negative

As mentioned earlier, this process uses a copper-clad board on which a light-sensitive chemical has been deposited. In order to make the chemical acid-resistant in the areas you want to be conductors, you must expose the board to light in these areas. To do this, you will need a negative of your circuit artwork, a piece of film in which the conductors are transparent and the background is black.

Start with your original layout. If the circuit is simple, you may want to stay with a life-size master. If the circuit is more complex, you may want to redraw the art two or even three times life size. Whichever size you choose, make your drawing very neat. If you draw, use drafting paper and a drafting pen. Do not use a pencil. If you are not used to preparing art for reproduction, do not try to use ballpoint or felt tip pens. Actually, the simplest approach to preparing your master is to buy adhesive-backed pads and lines from a mail order house or radio hobby store. These will produce the most professional results. Remember not to make sharp bends in your conductors.

When your positive master is complete, inspect it carefully. Check for dirt and smudges. Make sure all lines and edges are sharp. Then put your master in an envelope—**don't fold it**—and take it to a photo process laboratory. Ask a local printer to refer you to someone in the local area who can do **process** or **litho** photography. Ask the laboratory to make a line negative from your master. This shouldn't cost more than a couple of dollars. It's a service they do regularly for their customers, and you aren't asking anything unusual. Normally, you can get same-day or at least 24-hour service. What you will get back from the process laboratory will be a very high-contrast negative of your original artwork. Take it and follow the exposing, developing, and etching steps given below.

Sensitized Board and Mechanical Negative

There is an alternative to taking your finished master to a process laboratory and obtaining a photographic negative. You can, if you wish, prepare what is called a **mechanical** negative without using a photographic process.

You will need to go to an art supply store and ask for an appropriately sized sheet of a material called **Amberlith**. Another material, called **Rubylith,** will also do the job, but it is darker and it's harder to see the pattern through it. Either of these materials will appear to be simply a sheet of rather thick, transparent, tinted, plastic. Actually, the color comes from a very thin film on one surface of the plastic.

Tape your master artwork down on a hard surface and tape the Amberlith on top of it, **dull side up.** Use a sharp tool to outline each conductor and pad. You can use an X-acto knife or a pin, but the best tool is a pair of draftsman's dividers, or a bow compass with a point in each leg. With either of these tools, you can adjust the spacing of the points on the instrument to the width of the conductors you will be drawing and cut both sides of the conductor at once. You can use the instruments as you would an ordinary compass to scribe circles for pads (Fig. 2-5).

After you outline each conductor or pad, use a sharp implement to tease up a corner of your conductor and peel the

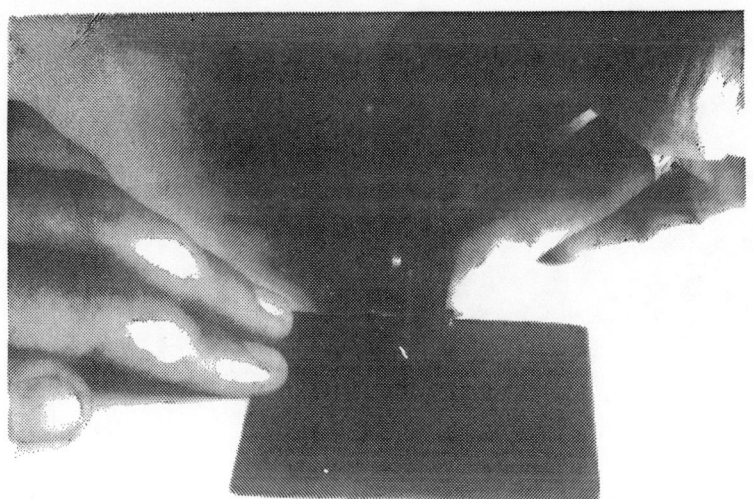

Fig. 2-5. You can use a bow compass to make parallel scribes for lines and round scribes for pads or Rubylith or Amberlith mechanical negatives.

film away from the transparent base material (Fig. 2-6). Be sure that you are removing the red film from the conductor area and not from the background. When you have finished, you will have a negative image of your circuit, with conductors clear and background red or amber.

The reason you can use this transparent red material in your negative instead of an opaque back material, as in a photographic negative, is that the photoresist chemical is not especially sensitive to light at the red end of the spectrum. Thus, to it, the red Rubylith appears as black as if it were opaque.

When you have completed your mechanical negative, you are ready for exposing, developing, and etching (see sections below).

Exposing

Before you take the light-sensitive copper-clad board out of its protective wrapper, you will have to darken the room somewhat. The photoresist isn't very sensitive, so you can have some light, but be sure there are no fluorescent lamps burning. You can use a 15-25W bulb, shaded so that no direct light falls on the board, 7 ft or more away. You can also use a red darkroom safelight.

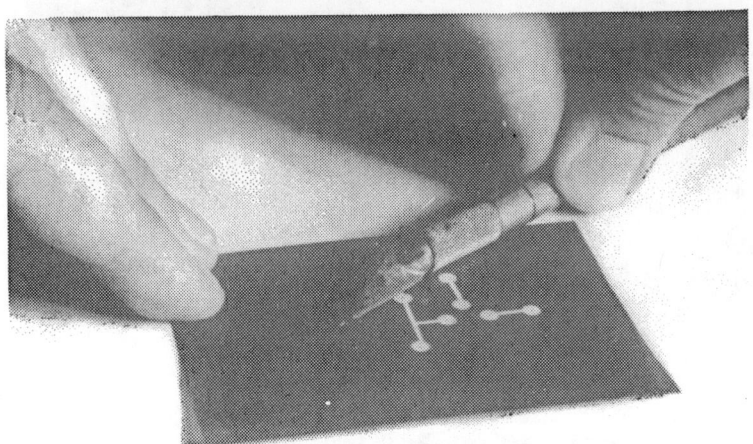

Fig. 2-6. After scribing your circuit, use sharp implement to tease up corner of film, and peel away in areas that are to be conductors in final circuit.

When you have the lights adjusted, sandwich the negative between the copper-clad board and a sheet of glass, and clamp the edges together. The copper side of the board should be against the dull side of the photographic negative or the shiny side of the mechanical negative. In other words, you should be able to see the circuit exactly as you wish it to appear in the final product, when you look at the sandwich. Of course, the negative and glass plate should be totally clean and free from dust specks, as each dust speck will show up as an imperfection in the finished circuit.

To make your exposure (Fig. 2-7), you can use either a standard 150W reflector lamp or a No. 2 photoflood bulb in a reflector assembly. If you use the 150W reflector lamp, make a 3½-minute exposure from a distance of 12 in. If you use the No. 2 photoflood bulb, make a 6-minute exposure at a distance of 10 in. Whichever bulb you use, plan your exposure so that your fingers and other things sensitive to extreme heat are not in the direct light of the bulb. After you make the exposure, keep the room darkened until you have finished the developing step.

Developing

Use an aluminum or glass tray for the developing process. A plastic tray may be dissolved by the developing chemical.

Place the exposed circuit board face up in the bottom of the tray and cover it to a depth of approximately ¼ in. with developer. You will notice that the developer is very volatile. It is best to use a sheet of glass to cover the tray to keep in the fumes. This holds down evaporation and is a desirable health precaution.

Gently agitate the developer for 2 minutes. At the end of this time, turn on the lights and tilt the tray so that all of the

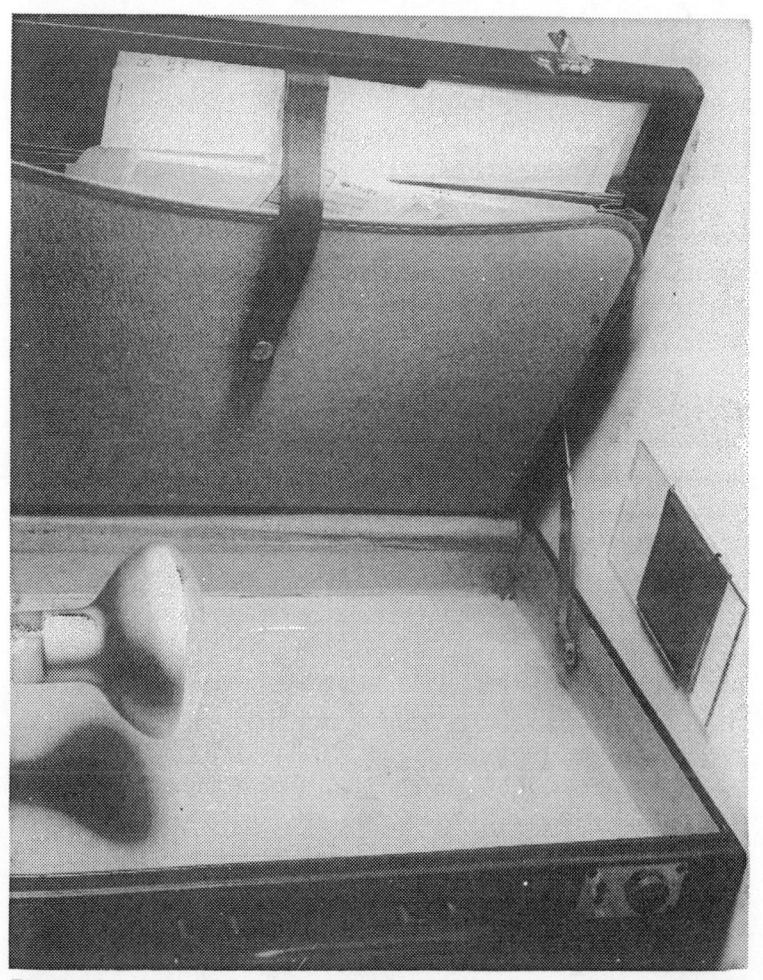

Fig. 2-7. Sandwich the sensitized copper-clad board and negative between glass plates and make exposure. Set things up so you do not have to hold the sandwich in your hands during exposure—these bulbs generate a lot of heat.

developer collects at the far end, and carefully remove the circuit board with fingertips or tweezers. Be very careful not to handle the surface of the board.

Allow the board to dry for 30 seconds to a minute. Do not wipe or shake the board, or even blow on it to hurry it up. The developer that remains in the tray can be reused once or possibly twice more before its potency is gone.

Etching

If you used an aluminum tray for developing the board, you cannot use it for etching. The etchant would attack it the same way it attacks the bare portions of the circuit board. For etching, you must use either a glass or a plastic tray.

Place the board in the tray and add the etchant until the board is covered. The etching will take anywhere from one to two hours, depending on the temperature of the solution. During this time, you must agitate the tank every 5 minutes or so.

As you watch the etching process, you will observe that the portions of the copper-clad board that are not protected with resist will take on a brownish color. If you pull the board out of the etchant at this point, you will notice that as the brown liquid gradually drains off the surface, the bare areas of the board show a color somewhat like the color of a very new penny. This is the natural color of copper, when it is not affected by a surface oxide. The etching process is not complete until there is no more of this rosy pink elemental copper on the board. Do not be fooled into taking the board out of the etch bath too early. Inspect the board periodically to find out how much of the copper remains. Be especially careful where there are narrow gaps between conductors. The etching is not finished until the only thing you can see is resist-covered conductor and dull brown circuit board.

After the etching has been completed, remove the board and rinse it under running water. You cannot reuse the etchant, so dispose of it, preferably down the commode, or, if in the sink drain, with a thorough water flush. Avoid handling the board with your fingers, since the etchant stains may cause irritation.

Finally, remove the resist from the conductors. This is essential to good soldering. Use plain steel wool, or steel wool and a solvent intended for removing resist.

Finishing the Board

Now that you have a board with a circuit, you must prepare it for the components you will be mounting on it. Each pad must be drilled in the center for the component lead that will go through it. There are special printed circuit board drills that you can buy for this purpose. They have a different angle to the cutting head than ordinary general-purpose drills. This helps them drill easily through the brittle circuit board without destructive stresses. For larger holes such as for mounting brackets and stud-mounted components, you will have to use your regular drill bits. Use moderate pressure and a slow speed. If the drill binds, do not try to get it restarted by running the speed control up and down. You will damage the drill motor if you do. Instead, work the board off the drill and get the drill going at medium speed without a load. Then apply the rotating drill bit to the hole, and it will cut right through.

Install the larger components first; then the smaller resistors, capacitors, diodes, and transistors. Push resistor and capacitor leads through their appropriate holes until the body of the component is flush against the board. Then bend the lead over against the board and clip off all except an ⅛ in. or so. Treat transistors in the same way, except leave about ½-¾ in. of lead on the top side of the board. This will permit you to attach heatsinks to the leads before soldering. Use a soldering iron with low heat.

In some manufacturing operations, components are soldered to printed circuit boards with a device called a **wave-soldering machine**. In this machine, there is actually a fountain of molten solder that is kept flowing at all times. Circuit boards go through the machine much like cars through a car wash. They first pass over a flux bath to clean all the leads and conductors, and then they pass over the solder fountain to solder all of the connections in one pass.

The usual home experimenter technique is to mount the finished circuit in some sort of case, using metal brackets and screws. Some more sophisticated projects use a technique borrowed from industry, called **edge card mounting**. In this technique, the circuit on the board includes a series of rectangular pads along one edge. These pads connect to the various inputs and outputs of the circuit. The board is intended to mate with a socket, called an **edge card connector**. This is a long connector with fingers that make contact with the pads on the circuit board. The board itself is the plug, and there is no need for a separate male connector.

The advantage of this is that a series of edge card connectors can be mounted adjacent to each other on a panel or another printed circuit board, called a **mother board,** and each stage of a device can be built separately and plugged into its own edge card connector. This aids in both design and troubleshooting. It aids in design because each stage can be modified independently on its own edge card connector, and comparisons can be made by just removing one design and substituting another. It aids troubleshooting because a faulty stage can be isolated quickly by simple substitution. As the technique becomes more common, you can expect to see more kits and magazine projects using edge cards.

METAL CHASSIS

Before the days of etched circuits and perforated boards, metal chassis construction was used for virtually all electronic assemblies. Today, its use is limited to equipment that requires it. For instance, most vacuum-tube circuits require periodic replacement of tubes. This places mechanical stresses on the surfaces used to mount tube sockets. Metal chassis are better equipped to handle resultant strains than are brittle circuit boards. Vacuum tubes are also great producers of heat, which is more efficiently conducted away and radiated by a metal chassis.

Vacuum tubes are not the only components that generate heat. Diodes and power transistors carrying large currents use the metal chassis on which they are mounted as heatsinks.

Heavy metal chassis are required to support the weight of transformers and chokes in high-voltage power supplies.

Construction Techniques

Before you do anything else to your metal chassis, cover it completely with some sort of material that you can write on—masking tape is ideal. This will give you an easy way to mark holes for cutting and will protect the finish of the metal from scratches caused by chips and burrs.

Next, determine the location of all your controls. Connectors, switches, and potentiometers will usually be mounted on the sides of the chassis or box. Variable capacitors can be mounted on either the side or top surface. All shafts and some mounting brackets of variable capacitors are electrically connected to the capacitor rotor, so plan to insulate the shaft and bracket if they are not to be at the same potential as the chassis.

Once your control positions are chosen, locate the positions for other large components such as transformers, chokes, filter capacitors, etc. Most of these will require nothing more than 9/32 in. holes for their mounting hardware. The easiest way to locate the holes is to position the component as it will be mounted in the final configuration, and then use its own mounting bracket as a template for marking the hole positions. Some of these holes can also be used as locations for solder lugs and terminal strips. If the component-mounting holes do not allow for enough conveniently located lugs and terminal strips, mark additional holes for them on the chassis.

Mark positions for tube socket holes on the top surface. Some sockets are fastened in place by means of spring clips; others have brackets with screw holes. If you are using the latter type, don't mark the holes for the bracket-mounting screws just yet. Wait until you have the socket holes punched or cut, and then, with the socket in the hole, use the bracket as a template.

Once you have the positions of all the major components laid out, you are ready to start drilling. If you do not have a variable-speed drill, use a center punch to locate the drill tip for each hole. If you have a variable-speed drill, you can skip the center punch and start at a low speed. For larger holes, ¼ in. and up, start with a smaller hole and work your way up.

The best technique for large round holes is to use a **socket punch** of the proper size. Drill a ⅜ in. hole first, to pass the center bolt of the punch, and then assemble the halves of the punch on either side of the chassis and tighten the bolt.

Chassis punches are best for making larger holes, but they are expensive. If you are careful, you can make an acceptable hole with a nibbler and round file. Nibble the hole just a little undersize, and then carefully enlarge it until its edges are smooth and it just fits the socket. Mark the locations of the socket-mounting holes and drill these as you have drilled the others.

When you have satisfied yourself that there are no more holes to be drilled, remove the protective paper and install the components, solder lugs, and terminal strips. Always use lockwashers between chassis and nut. These serve two purposes: They help to tighten the screw without need for a wrench on the nut, and they help to prevent the nut from loosening during vibration.

You may now wire up the chassis. Work from the schematic and proceed logically, a stage at a time. Any rf wiring should go directly from point to point. All other wiring should be routed so that it will be convenient to bundle the wires into cables. Cut leads on resistors and capacitors so that they can be installed with their edges parallel to edges of the chassis. It is best to wait until all wires that go to a particular point are installed and firmly bent over before soldering. If a wire is to pass through a hole in the chassis from top to bottom, use a rubber grommet or a feedthrough insulator.

GENERALLY APPLICABLE TECHNIQUES

There are some practices that are common to all four kinds of electronic construction. They deal with controls and shafts, and wire sizes and colors.

Controls and Shafts

Some potentiometers come with shafts already installed. In other cases, pots and shafts are sold separately, so that you can choose a round, half-round, slotted, or knurled shaft to suit the kind of knob you intend to use. These shaft styles are interchangeable, but once you insert the shaft into the pot and push it until it locks, you are committed. You cannot remove one of these shafts without destroying the pot.

On-off switches for the backs of volume control potentiometers are also sold separately. If you remove the gummed paper from the back of the pot, you will expose a small finger mounted on the pot rotor that is designed to mate with the switch actuator. Match the finger with the actuator and click the assembly together.

Obviously, you will not need the whole length of control shaft as provided by the manufacturer. Use your knobs themselves to find out exactly how much shaft you will need, and cut off the rest with a hacksaw.

Many of the pots readily available in radio parts stores are intended for printed circuit boards. You can tell these pots by the absence of holes for wires in their rather narrow solder terminals. Since these pots are intended to be mounted on brackets, only one nut is provided for their bushings. If you want to install one of these pots on a chassis or metal panel, you will find that the bushing sticks out far in front of the panel and prevents the knob from seating flush. Potentiometers intended specifically for panel mounting have two bushing nuts, one for either side of the panel. The rear nut can be adjusted for panel thickness, so that the front of the bushing does not protrude past the end of the front nut.

As already noted, variable-capacitor shafts are electrically connected to the rotor plates of the capacitor. Most designs take this into account and place one side of the capacitor, the rotor (always the curved element in schematics), at ground, or chassis potential. In some circuits, however, this is not possible. In these cases, it is necessary to insulate the shaft. Sometimes, all that is necessary is to provide a large hole for the shaft to pass through without touching any sides and a plastic or Bakelite knob. In other cases, for example, when you want to avoid side loads on the shaft, you will have to use a panel bearing coupled to the shaft through a flexible coupling. These are fairly uncommon parts, even in the best stocked retail stores, and you probably will only be able to get them through the mail order houses. These same sources can also provide flexible shaft extensions and right-angle drives, but you may go through an entire lifetime without ever needing these items.

Wire Sizes and Colors

The most common wire used in electronic projects is No. 22 stranded copper wire with plastic insulation. For most uses, stranded wire is preferred to solid wire, because solid wire breaks too easily when it is nicked.

For carrying heavy currents, No. 22 wire is too small. High-current applications require No. 14 or even heavier wire.

Some applications—for instance, coil winding—require very fine wire. Solid copper wire with an insulating coating of enamel is sold for this purpose. Number 28 wire is convenient, since it is fine (0.014 in.) yet reasonably strong. You will have to remove the enamel coating from the ends of this wire before you will be able to solder to it. Industrial users of enameled copper wire employ a paint-removing chemical to do this. You can achieve the same results, however, by scraping the ends of the wire with a sharp knife. Be sure to scrape off the enamel all the way around the circumference of the wire.

Of course, the color of the insulating jacket on the wire in your project will have no effect on the performance of the circuit. However, you can make the project look better and assist yourself in troubleshooting if you can afford to invest in several colors of wire. The colors in Table 2-1 represent standard practice for wiring different parts of a circuit.

Table 2-1. Standard Colors Used in Chassis Wiring.

Color	Use
Black	All grounds
Brown	Heaters or filaments off ground
Red	Positive dc supply voltage
Orange	Screen grids, base 2 of transistors, gate 2 of FETs
Yellow	Cathodes, emitters, sources
Green	Control grids, bases, FET gates
Blue	Plates, collectors, drains
Violet	Negative dc supply voltage
Gray	Ac supply leads
White	Bias and agc lines

Fig. 2-8. Multiple exposure of spinning silver dollar illustrates photographic effects possible with Multi-Strobe.

MULTI-STROBE PROJECT

This is a fun project that will add excitement to parties and provide some interesting effects for amateur photographers. It's a bright, flashing strobe lamp with a variable flash rate. It operates on house current.

At parties, with a moderate flash rate, the Multi-Strobe can add a very unusual "old-time movie" effect to fast dances. Seen in the light of the Multi-Strobe, dancers appear to move jerkily between exaggerated positions.

In photography, it is possible to use the Multi-Strobe to show a moving object in successive positions on a single frame of film. The photo of the spinning silver dollar in Fig. 2-8 was made in this way. The photo was taken at f/4 with the shutter held open for about 2 seconds. The Multi-Strobe was about 2 ft away. For this shot, the film used was Tri-X, with the development time extended to push the film's ASA value to 800. Most good camera shops will do this for you at slight expense.

The flash of the Multi-Strobe is very bright but of short duration. This can be used to allow you to see some very commonplace things in interesting new ways. For instance, when you look at a drop of water falling in ordinary light, the image is blurred and it is difficult to tell the shape of the drop. The flash of the Multi-Strobe is bright enough to let you see the drop, but short enough so that the image is not blurred. By

47

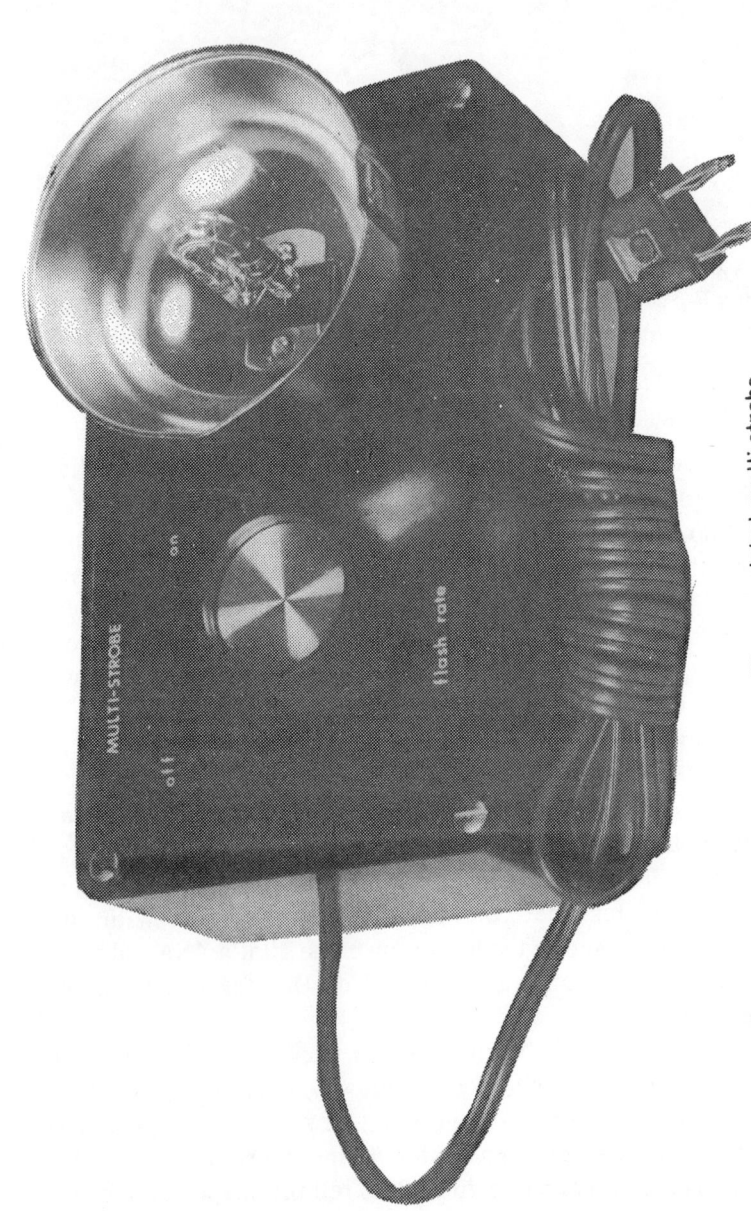

Fig. 2-9. The completed multi-strobe.

releasing drops from an ordinary eyedropper in front of the Multi-Strobe, you will be able to observe their true shape, which is spherical—not at all like a teardrop. Once you are used to using the Multi-Strobe, there are many other kinds of everyday phenomena you will find it interesting to study.

Principle of Operation

The Multi-Strobe is pictured in Fig. 2-9. The flash tube, the most important element of the strobe, is shown in Fig. 2-10. This U-shaped glass tube is filled with the inert gas **xenon**. There are three electrodes: an anode, or positive terminal; a cathode, or negative terminal; and a trigger band. In the photograph, a length of enameled copper wire has been soldered to the terminal for the trigger band and wound around the tube. This is a common practice with xenon flash tubes: It helps insure triggering.

Chemically, xenon is related to neon, which is very familiar in glow-discharge display tubes, and to argon, which is used in some ultraviolet light sources. The xenon flash tube differs from common neon and argon glow lamps in that the two electrodes are separated by a much greater distance. Thus, when a high voltage is applied across the xenon flash tube, it does not immediately ionize and glow.

A high-voltage spike—in the case of the Multi-Strobe, a 6 kV impulse—is used to trigger the tube into firing. This momentary high-voltage, low-current spike is obtained from a specially designed trigger transformer, much like a miniature version of the high-voltage flyback transformer in a television set.

Power Supply

Because the overall drain of the triggering circuit and flash tube is small, it is possible to use a voltage doubler power

Fig. 2-10. This xenon-filled flash tube is the heart of the Multi-Strobe.

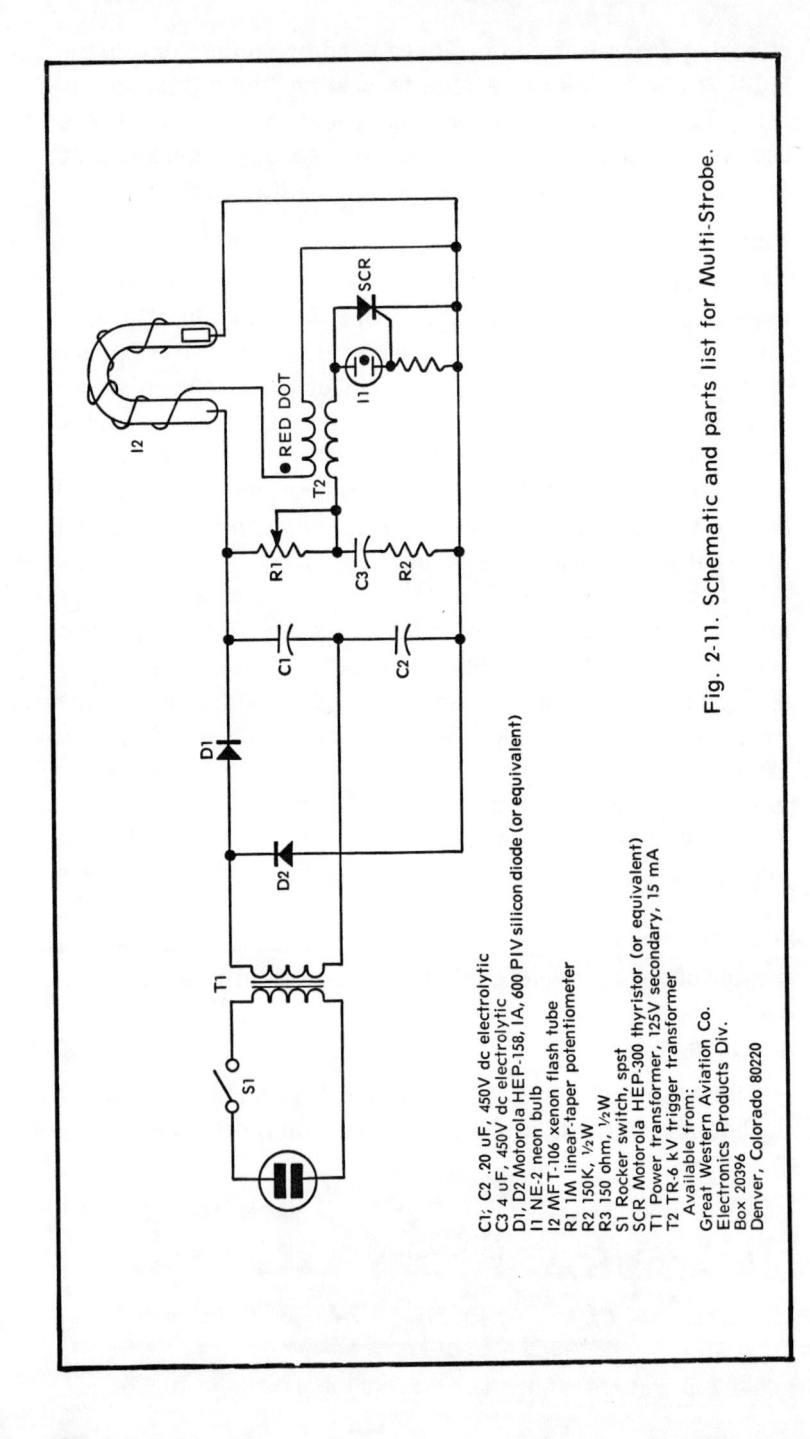

C1, C2 .20 uF, 450V dc electrolytic
C3 4 uF, 450V dc electrolytic
D1, D2 Motorola HEP-158, 1A, 600 PIV silicon diode (or equivalent)
I1 NE-2 neon bulb
I2 MFT-106 xenon flash tube
R1 1M linear-taper potentiometer
R2 150K, ½W
R3 150 ohm, ½W
S1 Rocker switch, spst
SCR Motorola HEP-300 thyristor (or equivalent)
T1 Power transformer, 125V secondary, 15 mA
T2 TR-6 kV trigger transformer
 Available from:
 Great Western Aviation Co.
 Electronics Products Div.
 Box 20396
 Denver, Colorado 80220

Fig. 2-11. Schematic and parts list for Multi-Strobe.

supply (see Fig. 2-11). This produces a relatively high voltage, around 300V, from an inexpensive transformer. This transformer is not really necessary for the functioning of the circuit. However, by isolating the components from the power lines, it provides an important safety feature.

Voltage Doubler

The basic voltage doubler circuit, isolated in Fig. 2-12, is a handy thing for the experimenter to remember, since it is a relatively cheap method of obtaining a high dc voltage. Its drawbacks are that it cannot supply very much power and cannot be used with a bleeder resistor.

Both of these drawbacks stem from the way the voltage doubler works. Each of the capacitors in the circuit provides half of the output voltage. Consider that the input wire going to the junction of the two capacitors is the neutral lead. Then we may consider that the other lead has an alternating voltage of plus and minus a certain voltage. During the positive half of the cycle, diode D2 is cut off and diode D1 conducts, charging capacitor C1. During the negative half of the cycle, D1 is cut off and D2 conducts, charging capacitor C2 in the same polarity as C1. If there is very little drain on the circuit, the voltage across the combination of capacitors is twice the peak voltage of the input signal. However, if there is a significant load on the circuit, the capacitors will be simultaneously

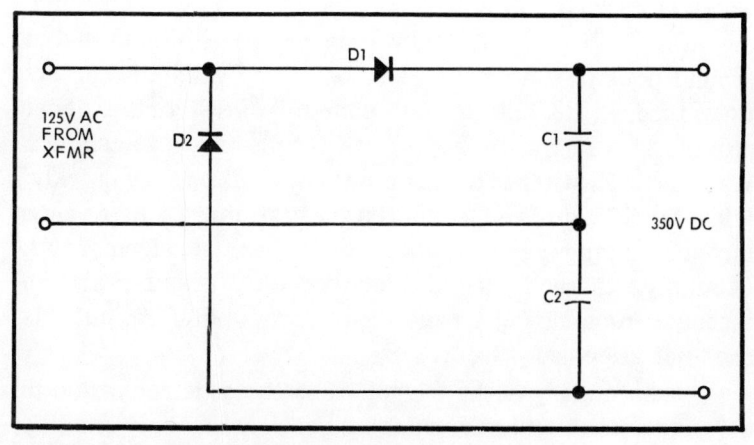

Fig. 2-12. Basic voltage doubler circuit.

charging and discharging (or, saying it another way, the diodes will be delivering power to the external load) and the voltage will be reduced.

In the case of the power supply in the Multi-Strobe, there is little load on the capacitors. The nominal output voltage of transformer T1 is 125V rms. This means that the peak voltage is 125 times 1.4, or 175V. Thus, the voltage doubler provides a 350V dc voltage at no load.

The use of a voltage doubler requires a note of caution. Because there is no bleeder resistor to discharge the capacitors when the switch is turned off, there is the possibility of a lingering charge on the capacitors that can produce an unpleasant (if not particularly dangerous) electrical shock. If it is necessary to open the case of the Multi-Strobe within a half-hour or so after it has been used, it is a good idea to use a length of wire to discharge the capacitors before handling any of the components.

Timing and Firing Circuit

The voltage from the power supply is applied to the RC network consisting of R1, R2, and C3. The rate at which C3 charges is determined by the setting of R1. Neon bulb I1 "sees" the whole voltage across C3, since until it fires, it does not allow any current to flow through R3. Neon bulb I1 does not conduct until the voltage across C3 reaches its firing voltage, which is approximately 90V. As long as I1 does not conduct, the gate of silicon controlled rectifier SCR1 is maintained at ground potential. As soon as I1 begins to conduct, current flows through R3, and the gate potential of SCR1 rises above ground. This forward-biases the SCR, allowing it to conduct. When an SCR switches to the **on** state, it turns on very quickly. When the SCR turns on, it discharges capacitor C3 all at once through the primary winding of trigger transformer T2. (Actually, there is some high-frequency ringing at the resonant frequency of C3 and the primary coil of T2, but this does not affect our circuit.)

The secondary of the trigger transformer is connected to the trigger coil on the xenon flash tube. Note that only one side of the secondary is connected to the trigger coil. The other side is connected to circuit ground. The sudden discharge of capacitor C3 through SCR1 produces a 6 kV spike across the secondary of T2, which fires the flash tube.

As soon as the voltage across the neon bulb drops below the extinguishing voltage of the bulb, it turns out and returns the gate of the SCR to ground potential. The SCR, however, does not cease conducting until almost all of the charge has been drained from C3. Then it too turns off. This allows the charging to start all over again, and the cycle is repeated as long as switch S1 is on.

Multi-Strobe Construction

Some of the components are mounted directly to a 6x3½x2 in. black plastic box, and some are mounted on an etched circuit board. The removable front panel of the box contains the flash rate potentiometer R1, on-off switch S1, and the reflector and socket for xenon tube I2. Transformer T1 is mounted to the bottom of the box.

Most of the things that could have been used as a reflector were fairly expensive. An ideal reflector, however, was found in the brightly plated cup of a 79-cent soup ladle, bought in a supermarket. The ladle's optical properties are entirely adequate for the relatively large flash area of the xenon tube.

The xenon tube has no plug as such, just three fairly stiff leads. A 3-conductor socket intended for a flat-pronged plug

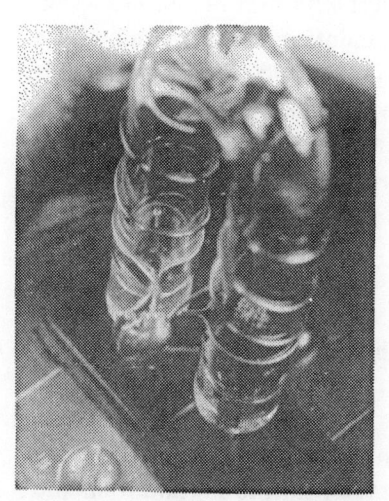

Fig. 2-13. Detail of soup-ladle reflector and 3-connector socket for xenon tube.

was selected to receive these leads. This required that a rectangular hole be nibbled in the bottom of the reflector and a corresponding hole be sawed in the front panel of the meter box. As shown in Fig. 2-13, the plug mounts by means of two 8-32 screws.

The on-off switch also requires a rectangular hole. The switch is mounted by inserting it into the hole until the side tabs lock in place. Its black plastic finish matches the box and makes a very attractive final product.

Figure 2-14 is a full-size pattern for the etched circuit board. The board is prepared as described in the chapter on printed circuits. There is some degree of nonuniformity from unit to unit in the positioning of the solder tabs on the trigger transformer. One technique for locating the position of the holes exactly is to place a sheet of carbon paper on a blank piece of paper and then to press down on the carbon paper with the transformer. this should leave four marks on the paper in the pattern of the solder terminals on the transformer. This pattern can then be transferred to the printed circuit.

The stud of the SCR is its anode, so it is important that good connection be made between the mounting hardware and the copper on the board. Use a couple of washers to limit the penetration of the stud, or you may crack the board when you mount it to the bottom of the meter box.

The circuit board mounts to the bottom of the box (Fig. 2-15). Metal or nonconducting standoff insulators can be used to mount the circuit board within the box; however, in this case, a simpler method was used. Holes were drilled through both circuit board and box, and 8-32 screws were passed through the holes with ¼ in. grommets used as spacers between the two surfaces. This is simple and compact, and at the same time provides vibration and mechanical shock protection.

For an orderly appearance, be sure to bundle leads into cables and lace them neatly as shown in the photos.

Multi-Strobe Operation

The operation is simple and straightforward. Plug the Multi-Strobe into a house receptacle, turn it on, and it should

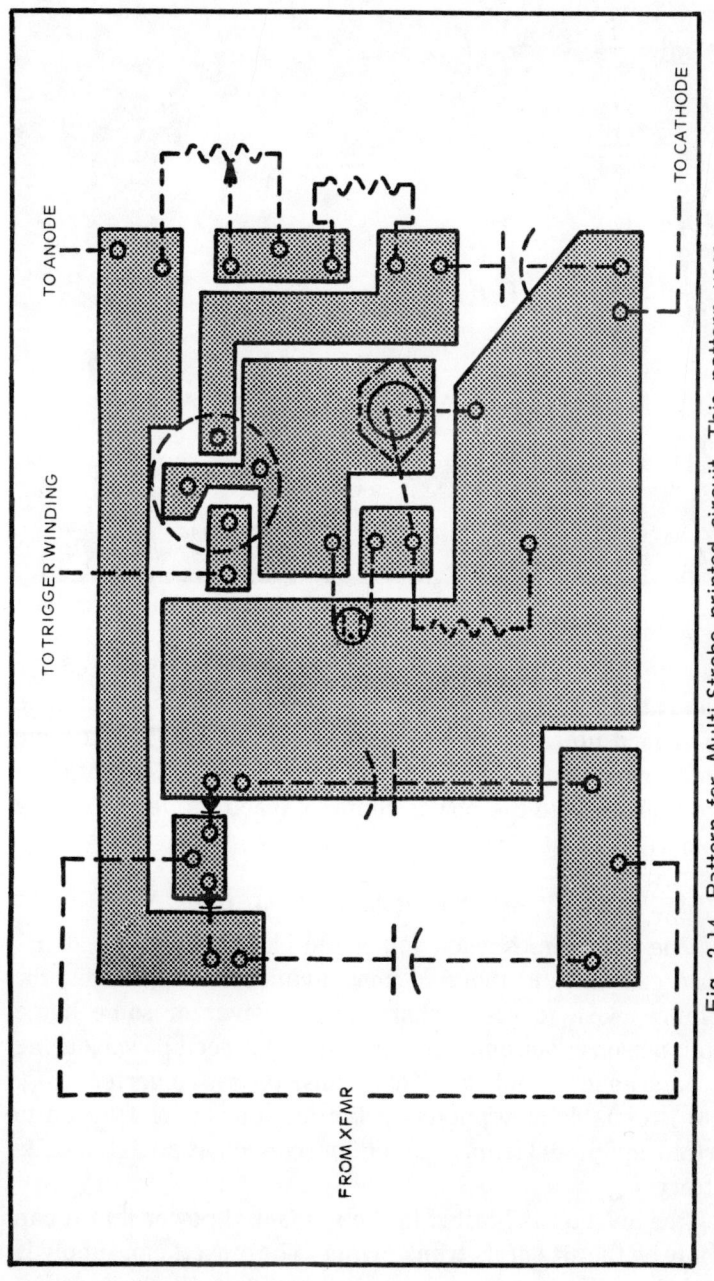

Fig. 2-14. Pattern for Multi-Strobe printed circuit. This pattern uses conductor blocks rather than pads and ribbon conductors.

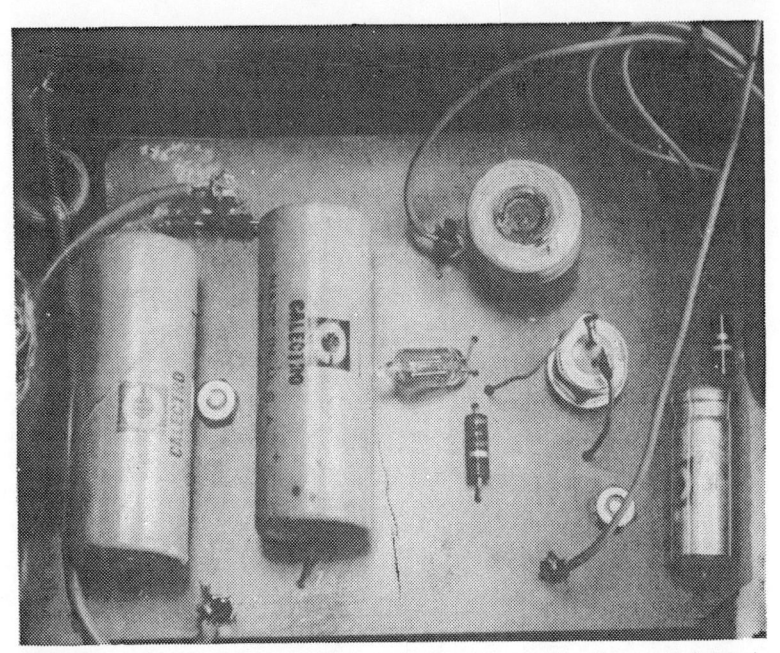

Fig. 2-15. Circuit board mounts to bottom of box; reflector, switch, and rate potentiometer mount on plastic cover.

begin to flash immediately. The potentiometer selected for flash rate control should vary the flash rate from about once per second up to around 10 flashes per second. If you have made any mistakes and the Multi-Strobe does not work, be sure to discharge the power supply capacitors before poking around with your fingers.

12V DC TO 110V AC INVERTER PROJECT

There are times when you would like to have 110V, 60 Hz house current and there is none available. You might, for instance, wish to use a shortwave receiver or some home appliance on a camping trip, or you might need power during an emergency blackout. The transistorized inverter (Fig. 2-16) is capable of supplying a limited amount of 110V ac to certain equipment from a 12V dc source such as an automobile battery.

The inverter is limited in the amount of power that it can supply by the size of its transformer. The most it can supply is 25W, given the transformer in the parts list. The transistors are good for 50W each at the voltage levels used, however; so if you wanted to use a transformer with a higher current rating and install it in a larger box, you could expect to supply up to 100W of output power. A second limitation is the magnitude of the load power factor that the inverter can accommodate. This means that electric motors will not work when connected to the inverter. The inverter is well suited for use with electronic equipment, however, provided power requirements are kept in mind.

The inverter uses a transistorized multivibrator to "chop" the 12V input signal into a square wave (see Fig. 2-17). The transformer is then able to step up this alternating voltage to 110V. This is the same thing that all transistorized inverters do. This particular inverter circuit differs from most in some important particulars, though. Most conventional transistorized inverters employ a saturable-core transformer with additional windings to make the multivibrator function. These transformers are available from several manufacturers at fairly stiff prices. Our inverter uses an ordinary 24V filament transformer, which lowers the price somewhat. The key to this simplified design is the use of **tantalum** electrolytic capacitors in the feedback circuit in the multivibrator.

Fig. 2-16. Transistorized 12V dc to 110V ac inverter.

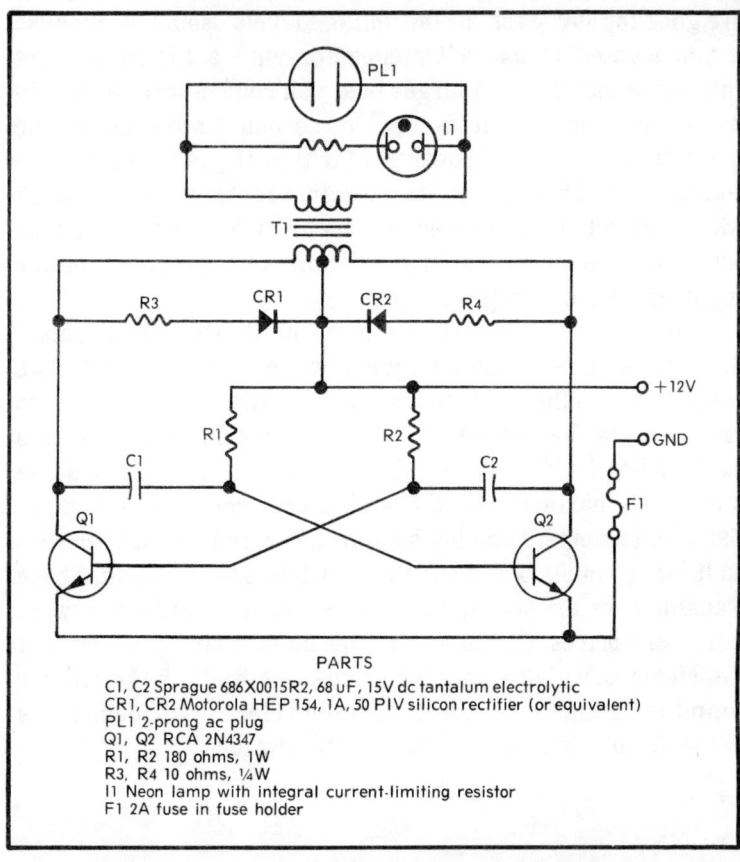

Fig. 2-17. Schematic and parts list for transistorized inverter.

PARTS
C1, C2 Sprague 686X0015R2, 68 uF, 15V dc tantalum electrolytic
CR1, CR2 Motorola HEP 154, 1A, 50 PIV silicon rectifier (or equivalent)
PL1 2-prong ac plug
Q1, Q2 RCA 2N4347
R1, R2 180 ohms, 1W
R3, R4 10 ohms, ¼W
I1 Neon lamp with integral current-limiting resistor
F1 2A fuse in fuse holder

To obtain an output frequency of approximately 60 Hz, some large values of capacitance are necessary. These values of capacitance are only available in electrolytic capacitors. Conventional electrolytic capacitors, however, are not well suited to this kind of application. They are not purely capacitive. They have, in effect, a resistance across their capacitance that diminishes their charge-holding ability. In power supply and audio coupling applications, this is not a drawback, but in timing applications, it is. Electrolytic capacitors made with the metal tantalum are a rather recent development. They are substantially better than older types of electrolytic capacitors in terms of dissipation, and they can be used successfully in timing circuits such as this inverter.

Inverter Circuit

Like the astable multivibrators used as clocks and tone generators in many low-power circuits, the inverter uses RC coupling to feed a signal back and forth between two common-emitter amplifiers. In Fig. 2-18, the inverter circuit has been redrawn a little more simply, with the transformer secondary omitted so that the use of the transformer primary can be better shown. As you can see, half of the primary—from left end to the centertap—is used as the load for Q1, and the other half—from centertap to right end—is used as the load for Q2.

Let's take a look at what happens when we apply power to the multivibrator. Initially, current flows through both transistors. It is unlikely, however, that the two transistors will be so perfectly matched that the current through each will be identical. Let us say, for argument, that more current flows through Q1 to begin with. As more current flows through the transistor, the voltage drop across its load increases and the collector voltage drops. Since the collector of Q1 is connected to the base of Q2 through capacitor C1, this dropping collector voltage causes a corresponding drop in the base voltage, and consequently, bias current of Q2. With less bias, Q2 conducts less and its collector voltage increases as current through its load decreases. This increasing collector voltage on Q2 is

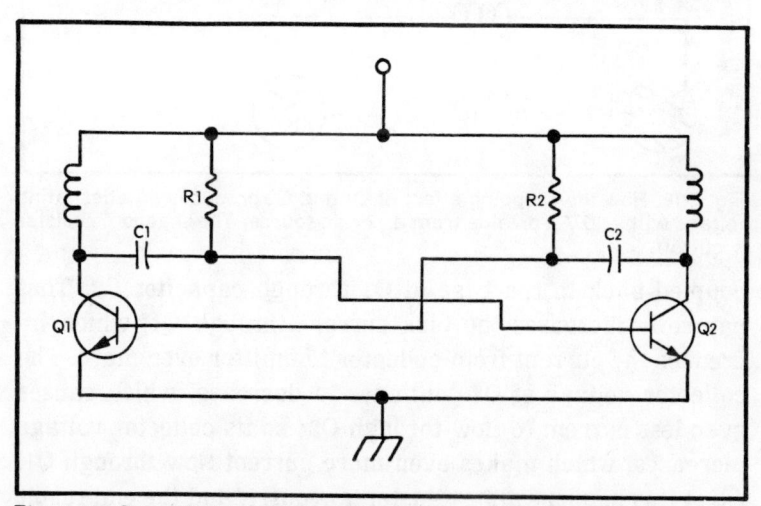

Fig. 2-18. Simplified inverter schematic showing how load formed by transformer winding is distributed between Q1 and Q2.

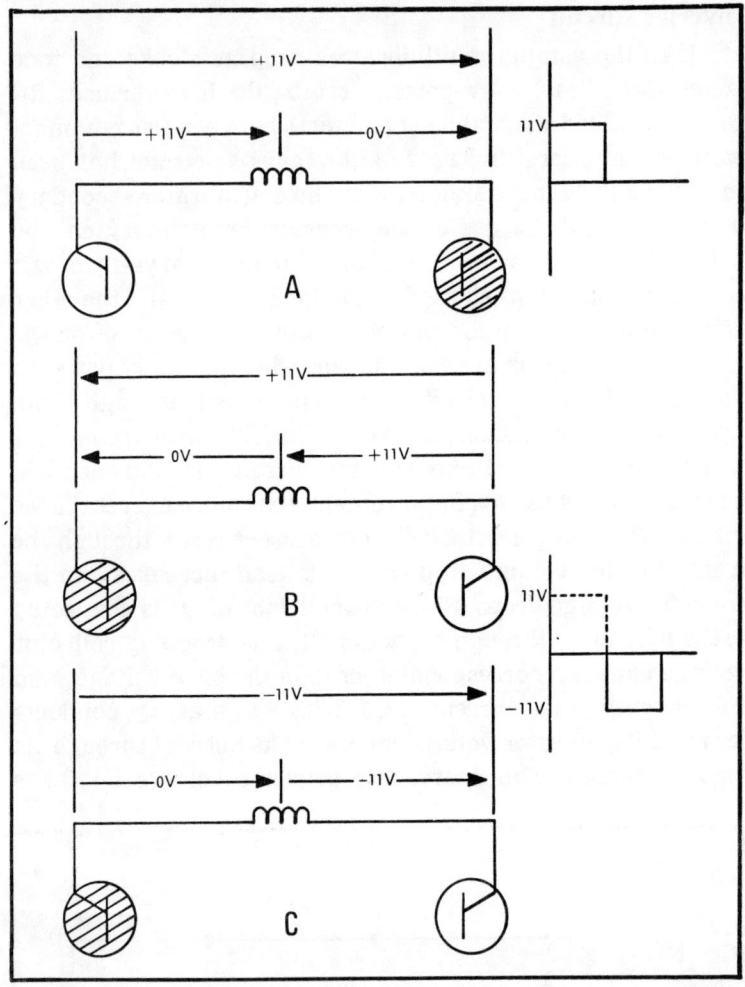

Fig. 2-19. How the chopping effect of Q1 and Q2 produces an alternating voltage with a 22V p-p value from a 12V dc source. The shaded transistor is cut off.

coupled back to the base of Q1 through capacitor C2. That naturally increases the bias current through Q1, which increases the current from collector to emitter even more. The collector voltage of Q2 continues to decrease, which causes even less current to flow through Q2; so its collector voltage increases, which makes even more current flow through Q1. All of this happens almost instantaneously, and the end result is that Q1 is fully saturated and Q2 is fully cut off.

This condition does not exist permanently, because of the effect of R1 and C1. With the collector end of C1 at essentially ground potential—thanks to the saturated state of Q1—and one end of resistor R1 connected to the +12V supply, C1 begins to charge up at a rate determined by its capacitance and the resistance of R1.

When the voltage across C1 reaches a point that allows Q2 to be forward-biased and begin conducting, the current through Q2 causes a voltage drop across its load that, in turn, causes a reduction in the collector voltage of Q2. As in the case of the initial conditions, this drop in collector voltage on one transistor is coupled to the base of the other. Eventually, this has the effect of reducing the current through Q1 a little. With less current flowing through the collector load, the collector voltage of Q1 rises, causing an additional voltage increase at the base of Q2. Once again, the feedback conditions create a rapid change of state; Q1 is almost instantly cut off and Q2 is almost instantly driven into saturation.

What effect does this oscillation have on the transformer? Figure 2-19 will help make it clear. In Fig. 2-19A, Q1 is saturated and Q2 is cut off. Since no current flows through the right half of the transformer primary winding, point A is at a potential of 12V. At saturation, the voltage drop across Q1 is only 1V, so point B is at a potential of 1V and the potential from B to A is +11V. When Q1 shuts off and Q2 is driven to saturation, the situation reverses: Point B is at +12V and point A is at +1V (Fig. 2-19B). In this case, the voltage from A to B is +11V, but another way of saying this is that the voltage from B to A is —11V, as indicated in Fig. 2-19C.

Effectively, then, the peak-to-peak voltage across the transformer primary is 22V, with half of the cycle at +11V and half at —11V. The transformer, which was intended to transform 120V ac to 24V ac, is a bilateral device, so it is capable of stepping the 22V ac to 110V ac, which is what it does.

Additional Considerations

The above analysis is accurate, as far as it goes, but it neglects two important contributions of the transformer.

Fig. 2-20. Parts of the inverter are assembled on top and sides of box. Point-to-point wiring techniques are used. This is an example of metal chassis construction discussed earlier in chapter.

Both contributions of the transformer affect the rate of change of the voltage across it. Any inductance resists rapid changes in voltage. The transformer, with its many turns of wire and its highly permeable iron core, possesses several henrys of inductance. The first effect this has is to slow down the rate of change of primary voltage so that the waveform of the voltage appears less like a square wave and more like a sine wave. Another way of looking at this effect is to say that the inductance blocks the higher frequencies present in the square wave. The second effect is the back emf that the coil generates in opposition to the rapid voltage changes. This takes the form of very high-voltage spikes that appear between the centertap and ends of the primary winding. These spikes are opposite in polarity to the normal voltage across each half of the winding. On the secondary (output) side, the spikes appear as large impulses of voltage.

These can seriously damage semiconductor devices in loads connected to the inverter. To get rid of the spikes, we include CR1 and CR2 in the design. These diodes are placed between the centertap and ends of the primary to short circuit the spikes and keep them from appearing in the output.

Resistors R3 and R4 are intended to limit the current through CR1 and CR2 to levels that will not damage the diodes.

Inverter Construction

All of the parts are assembled on the top and sides of a 3x4x5 in. aluminum box (Fig. 2-20). The transformer used in the inverter in the photos is smaller than the transformer recommended in the parts list, but you will encounter no trouble fitting the larger unit into the box.

A proper fuse is essential. Use a fuse with double the rating of the transformer. For example, the recommended transformer in the parts list is rated at 1A at 24V. Since the fuse is located in the 12V input line, a 2A fuse is called for.

It's a good idea to have a neon lamp to indicate that you have 110V at your output plug. The unit shown is self-contained. You can also use an NE-51 lamp in a pilot lamp socket, but be sure to include a 50K resistor in series with one of the leads to limit current.

Figure 2-21 shows how to install the power transistors. The outer shell of these transistors is the collector connection, so

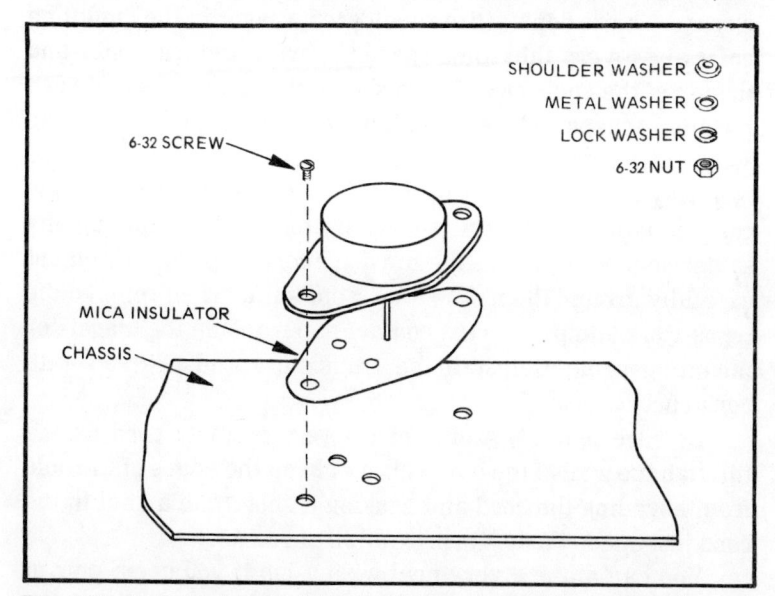

Fig. 2-21. Method of assembling switching transistors to provide electrical insulation from chassis. Mica insulator and shoulder washer come with transistor.

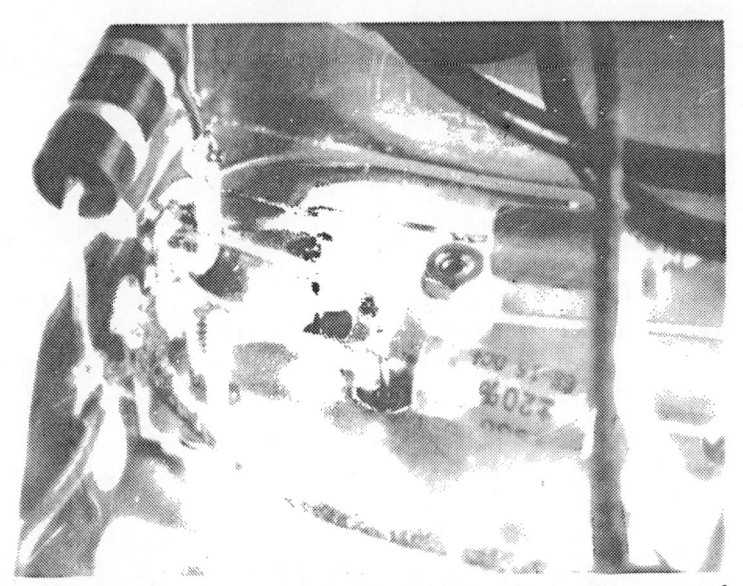

Fig. 2-22. By connecting a 4-pole terminal strip to one mounting screw of each transistor, we automatically provide a solder lug for the collectors and enough additional soldering lugs to make for neat construction.

the body of the transistor must be insulated from the enclosure. This is done with a shouldered washer in the mounting holes and a very thin mica spacer between the transistor and the top of the enclosure. There shouldn't be anything between the mica spacer and the top of the box, so you will have to take some extra care if you use contact paper to cover the box. The best way to make a neat job of it is to cover the box with the contact paper, then install the transistors without making any solder connections. Use the mica spacers as patterns and cut carefully around them with an X-acto knife. Then remove the transistors and peel up the contact paper inside the transistor mounting area. Reinstall the transistors and proceed with construction.

Be sure to use a grommet where the power cord passes through the wall of the box. This will keep the edges of the hole from abrading the cord and causing a short. Tie a knot in the cord just inside the box to provide strain relief.

You can make a very neat wiring job if you use a pair of 4-lug terminal strips as shown in Fig. 2-22. One of the terminals becomes the collector lead for each transistor, and it becomes possible to make a very symmetrical pattern of connections. Remember to use heatsinks when you solder to the emitter and base leads of the transistors.

TRIAC CONTROLLER PROJECT

There are a lot of circuits for motor-speed controls and light dimmers using SCRs and triacs. This is one of the best. It uses a triac rather than an SCR, so it provides full-wave control and excellent efficiency. It employs a dual time-constant control which minimizes hysteresis and permits very small turn-on angles. It uses a triac with an electrically insulated mounting tab, eliminating the necessity for a separate heatsink insulated from the metal enclosure. Finally, the triac is protected from commutating voltage effects by an RC circuit.

This type of controller is especially useful for the home experimenter as means of controlling the speed of electric drills. It is also ideal for controlling the heat output of soldering irons, and it can be used by photographers to control the brightness of photoflood lamps. In the latter use, it is best not to shoot color with the lamp at anything but its brightest voltage. However, it is very useful in setting up shots, since the lamp's effects on shadows and highlights can be studied with the lamp at a lower intensity. This makes it easier on the subjects—and on the photographer too, if the subjects are children.

The entire project fits into a 2¼x2¼x4 in. aluminum box (Fig. 2-23). The male plug is located right on the side of the box, eliminating the need for a cord. If it is inconvenient to use the controller right at the outlet, an inexpensive extension cord can be used to provide flexibility. The female ac connector is located on the end of the box opposite the male. The control potentiometer and fuse holder are mounted on the top of the box. A 6A fuse was selected to provide a safety factor for protecting the 8A triac (Fig. 2-24).

There are two terminal strips for mounting the components. The three terminals of the triac are spread out and soldered to three insulated terminals of one strip (Fig. 2-25). This provides a solid base for soldering the other components.

Fig. 2-23. Triac motor-speed controller.

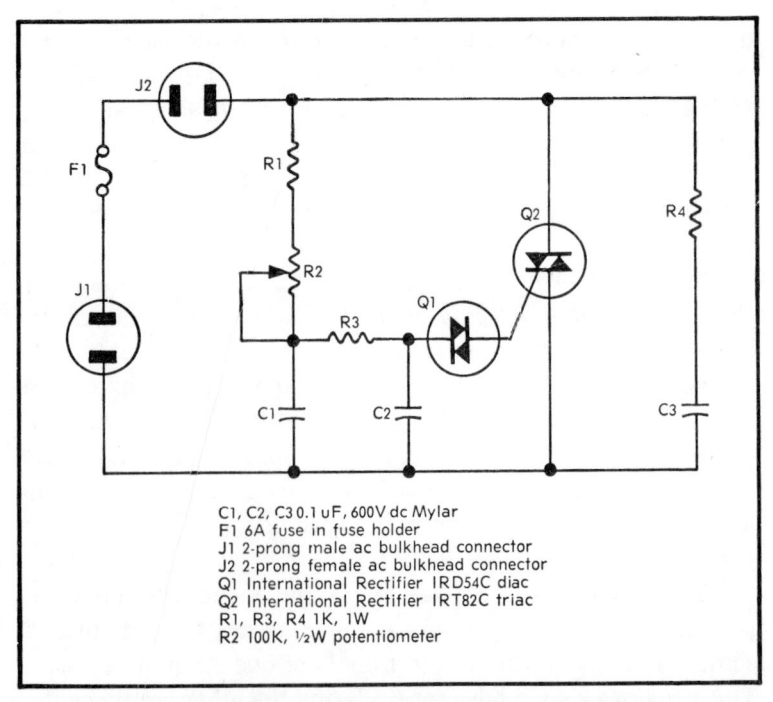

C1, C2, C3 0.1 uF, 600V dc Mylar
F1 6A fuse in fuse holder
J1 2-prong male ac bulkhead connector
J2 2-prong female ac bulkhead connector
Q1 International Rectifier IRD54C diac
Q2 International Rectifier IRT82C triac
R1, R3, R4 1K, 1W
R2 100K, ½W potentiometer

Fig. 2-24. Triac controller schematic and parts list.

With a terminal strip on either side of the box, it is easy to mount the resistors and capacitors neatly.

Speed Controller Circuit

A triac is like a pair of SCRs connected back-to-back in parallel, with a common gate connection. An SCR is a kind of a diode that does not conduct in the forward direction until the voltage across it exceeds a certain amount. Once it begins to conduct, it will continue to do so even though the voltage across it is considerably reduced. The SCR has a third electrode, called a **gate**. If a current is applied to the gate, the SCR will begin to conduct at a lower voltage. You can see that if there is a voltage across the SCR just a little less than the firing voltage, the SCR will behave like an open circuit until a current is applied to its gate, at which point the SCR will behave like a short circuit until the applied voltage falls to almost zero.

As we said, the SCR is like a diode. In the reverse direction, it always behaves like an open circuit. In ac control

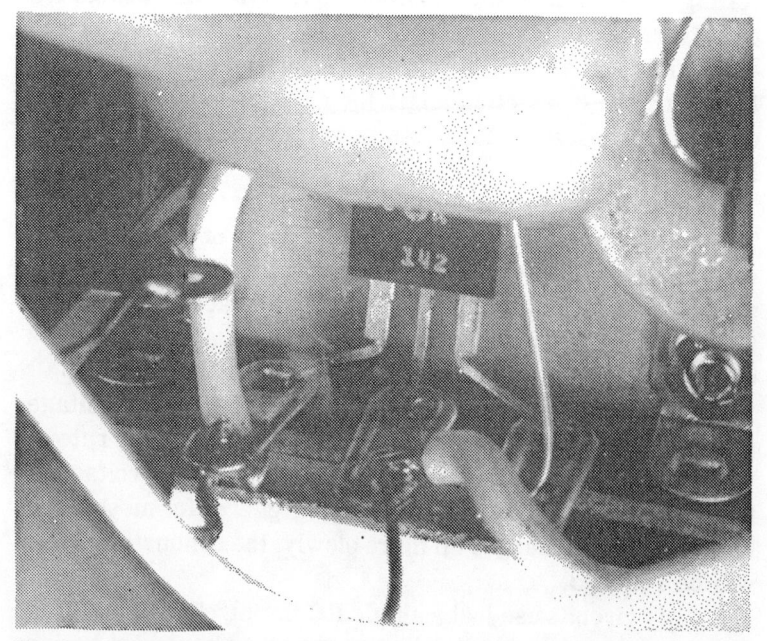

Fig. 2-25. The leads from the triac are spread out and soldered to the lugs of a 3-lug terminal strip.

situations, this means that only half of the sine wave is used to power the device being controlled. How can we use both halves of the sine wave? We can use a triac. The gate of a triac controls the firing voltage in both directions.

How is this effect used to control the power going to a load? It's a question of timing. The gate is used to turn the triac on during a part of each half-cycle. If the time is very short, the power reaching the load is small. As the time increases, so does the power. Engineers find it convenient to speak of this effect in terms of **conduction angle**. In one complete cycle of a sine wave, there are 360 electrical degrees. Depending on the setting of the control potentiometer and the values of the other circuit elements, the triac will conduct during part of the positive half of the cycle (say, from 65 to 179 degrees) and during an equal part of the negative half of the cycle (say, from 245 to 359 degrees). Under these conditions, the conduction angle is 114 degrees during each half-cycle, and quite a bit of power is getting to the load. If we changed the setting of the control potentiometer so that the triac conducted only from 150 to 179 degrees during the positive half of the cycle and from 230 to 359 degrees during the negative half of the cycle, the conduction angle for each half-cycle would be only 29 degrees, and much less power would be reaching the load.

How does the timing circuit control the conduction angle? The timing circuit consists of a dual RC network and a device called a **diac**. The diac is like a triac without a gate. It does not conduct in either direction until the voltage across it exceeds about 30V. At this point it begins to conduct, and it continues conducting until the voltage across it falls almost to zero.

In the circuit, the voltage across the diac is the voltage across C2. The values of C1, R1, R2, and R3 control the rate at which C2 charges. If C2 charges up to the breakover voltage of the diac very quickly, the triac will be triggered for most of the half-cycle. If C2 charges up more slowly, the conduction angle will be reduced.

Many circuits use just a single RC timing circuit, omitting R3 and C2, but by using these, we are able to control the triac down to very small conduction angles and to reduce an effect that engineers call **hysteresis**. In the case of lamp controllers, "hysteresis" refers to the difference in settings of the control potentiometer between the position where the lamp goes out and the position where it lights again.

The series resistor and capacitor across the triac is another circuit feature that does not appear in all lamp dimmer-motor controller circuits. Its purpose is to help maintain control when inductive loads such as motors are connected to the circuit. What can happen without the protective circuit is this: The inductive nature of the load causes the current in the circuit to lag behind the voltage. Thus at some point after the current through the triac has passed through zero and the device is cut off, there is an applied voltage across the device. If the rate at which this **commutating voltage** rises is fast enough, it may trigger the triac prematurely, causing loss of control. Thus, C3 and R4 actually form a simple high-pass filter that slows down the rate of increase of the commutating voltage to a rate that the triac can easily handle.

3

Finishing Touches

The appearance of the outside of your project—its "packaging"—ultimately spells the difference between a piece of electronic gear you are pleased to use and proud to show off, and a monstrosity you apologize for or hide under your workbench. There are a number of techniques to use and considerations to take into account that can make your projects look factory-made. Most of these tricks involve only a little extra effort and expense, and all of them will repay you manyfold by giving you a project with that "something extra." In this chapter, we will talk about selecting cases and enclosures; laying out a "scientific" and attractive control panel; selecting knobs, switches, and indicators; dressing up your panels with professional-appearing labeling; and making special scales for meters.

CASES AND ENCLOSURES

There are enough different kinds of cases and enclosures to fill up several pages in the average electronics catalog. If you are familiar with the various kinds of metal and plastic boxes, you will be better able to select just the right package for a particular project. There are more kinds of metal boxes than there are plastic ones, so we will look at them first.

The first thing to consider about metal enclosures is the materials of which they are made. You will frequently find the same box available in either aluminum or steel. For some kinds of shielding, steel is superior, and it certainly makes a more rigid structure. However, it is difficult to machine without elaborate equipment. Stick to aluminum unless you have a machine shop. In one of my first electronic projects, I slavishly followed a design for a ham transmitter published in a handbook. This design included a steel cabinet. It took the better part of a month just to hack a ragged hole for a meter in the front panel. One easy way to distinguish between steel and aluminum hardware, if there is no label and you can't tell by hefting it, is the finish applied by the manufacturer. Steel cabinets are almost always coated with black wrinkle-finish paint—like something out of a 1930s horror movie. Aluminum cabinets are more often finished in gray hammertone or left "natural." Some manufacturers are now offering cabinets finished in colored textured paint. These are all aluminum.

Miniboxes

The most useful style of metal enclosure for projects that do not require a lot of tuning and adjustments is the 2-piece standard minibox. This is an economical design, because it is easily stamped and bent by the manufacturer and because it offers a complete 6-sided enclosure with just two pieces (Fig. 3-1). One of the U-shaped pieces has side flanges that permit it to mate with the other U-shaped piece, and the flanges make the piece very rigid. Generally, all parts are mounted on the top panel and two sides of this enclosure, and the plain piece is

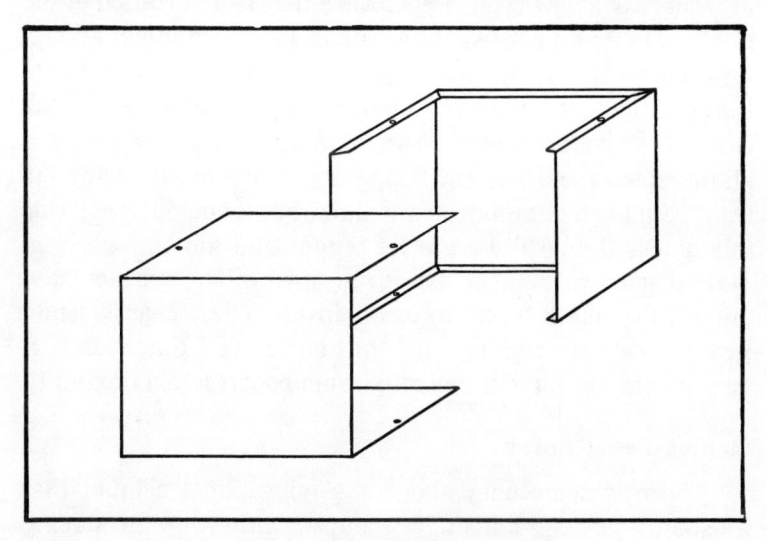

Fig. 3-1. The standard minibox.

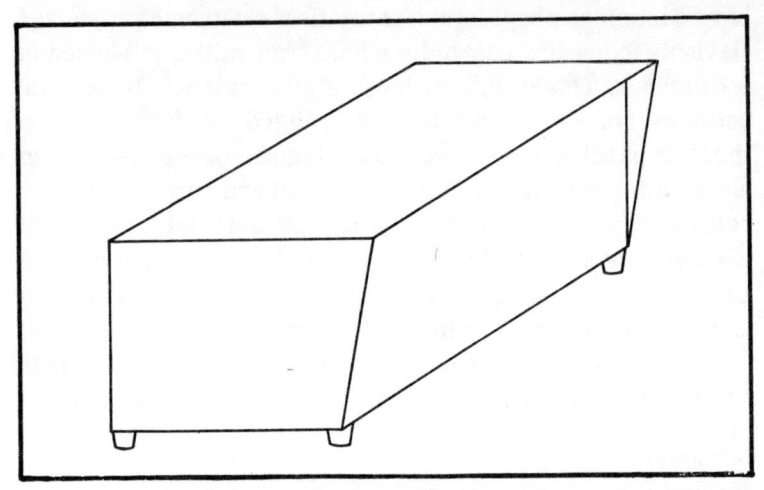

Fig. 3-2. The cowled box is more dramatic in appearance than the plain minibox, but may omit side flanges, providing less rigidity.

left to serve as a cover. Although the box by itself may look a little austere, its visual impact can be improved by painting the two halves contrasting colors. Alternatively, wood-grained contact paper can make the box look like a piece of cabinetry.

Cowled Miniboxes

There is a new type of enclosure that is a variation of the minibox in that it consists of a pair of U-shaped metal stampings; however, its appearance is much snappier. This is due to the cowl that shades the front panel (Fig. 3-2) and the way the case sits on rubber feet. The advantages of this enclosure are all in appearance. The cowl suggests that it might shade the front panel and eliminate some glare, but it couldn't really do this unless the cowl were a lot bigger and the box were installed somewhere near eye level. Most of these boxes have snug-fitting shallow chassis on the inside. These chassis lend a great deal of rigidity to the structure. Since they're removable, the builder has maximum construction flexibility.

Sloping-Panel Boxes

There is something about a sloping panel cabinet that whispers "test equipment." Perhaps it's the way the sloping panel presents its information. There is no doubt that the information is there to be read precisely and accurately. Sloping-panel boxes come with and without prepunched holes for meters. The kind with the prepunched holes are handy—if you want your meter smack in the middle of the panel. If your esthetic sense demands that the meter go to one side or the other, you will have to go to work with nibbler and file. Since sloping-panel boxes are not symmetrical, there are two ways to use them. If you have a lot of controls and switches you want to arrange in a neat row, orient the box as in Fig. 3-3A. On the other hand, if you have binding posts or sockets to arrange in order, you might try swapping bottom and back, as in Fig. 3-3B.

Utility Boxes

The utility box of Fig. 3-4 requires some welds, so it is more difficult for the manufacturer to make and, consequently, more expensive for you to buy. This is the kind of construction you will need, however, if you are using the conventional chassis-and-faceplate type of construction common to most vacuum-tube projects. In this kind of construction, you'll find several controls, pilot lights, switches, or connectors extending through both the panel and the front of

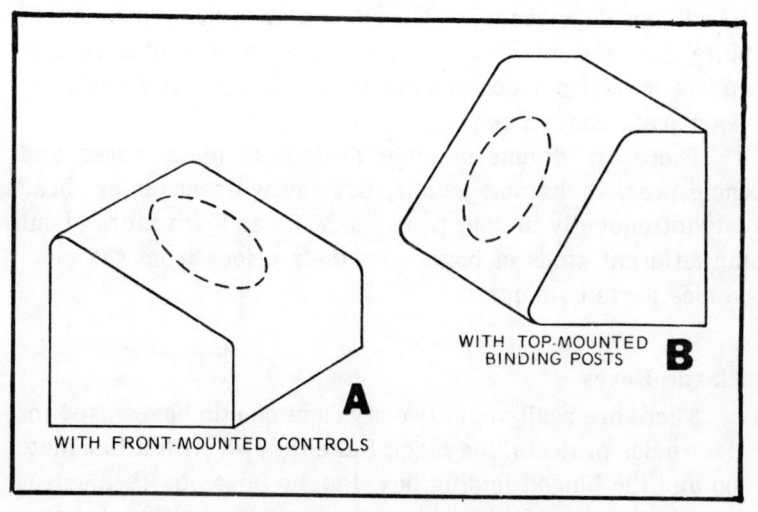

Fig. 3-3. Two ways of using a sloping-panel box.

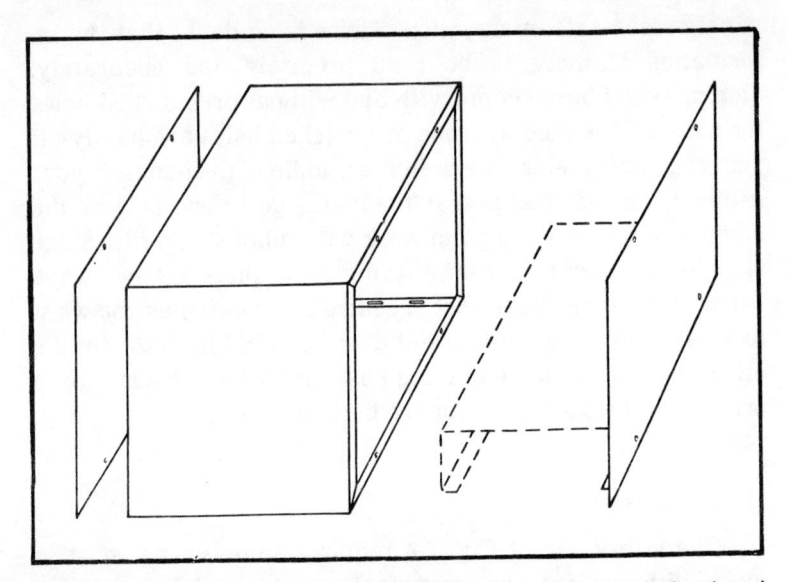

Fig. 3-4. The standard utility box provides removable panels front and back, creating a combination of rigidity and easy access. This box is often used with a separate chassis (dashed) for construction of projects using vacuum tubes.

the chassis, and these are used to fasten the two together. The faceplate is then attached to the cabinet with screws. This type of cabinet has the advantage of offering easy access to the circuits inside, since it is a simple matter to remove the backplate. Like the minibox, the utility box has a utilitarian look and benefits from contrasting paint on front and sides, or wood-grain contact paper.

There are dozens of other designs of metal cases and enclosures, racks and panels, but you will encounter these only infrequently in your projects. You can learn more about the different kinds of boxes and their prices from the electronics parts catalogs.

Plastic Boxes

There are really only two styles of plastic boxes used for electronics projects, the black Bakelite dish with a flat plate top and the hinged folding box. Plastic boxes by themselves provide no rf shielding, but they are favorites of home builders, because they look neat and are easier to drill and cut than are metal boxes.

Plastic boxes should be drilled with a hand drill or an electric drill adjusted to a slow speed. High speeds can melt some of the plastics used, forming hard burrs alongside holes. Some thinner plastic panels can be nibbled, although the edges will be rough and need to be filed. It is best, however, to use an X-acto keyhole razor saw after first drilling a 9/16 in. starting hole. The saw cut should also be dressed with a file, but you will find much less filing will be required for the saw cut than for the nibbled edge.

Transparent Plastic Boxes

In the days when transistors were still a novelty and the cheapest units cost several dollars, there were many, many projects published each month for one- and two-transistor pocket AM receivers and audio oscillators. It seemed that the hallmark of all of these projects was that they were built to fit into hinged plastic boxes bought at the five and dime. It is possible that overuse of these clear plastic boxes in those days has discouraged their use today, since projects using them are rarely published. Questions of style aside, these boxes still have their uses, especially in more frivolous projects, such as the Randomizer in this book. Here, half the fun comes from being able to see the simplicity and compactness of the inner workings through the transparent sides.

The publisher of this book manufactures a wide variety of electronic kits that employ plastic boxes as chassis. The kits are well designed and worth recommending, and they'll provide valuable wiring experience for the newcomer. In TAB's kits, plastic chassis cutting is accomplished by using a soldering iron as a melting "saw" and "drill."

Black Bakelite Boxes

The black Bakelite dish is the kind of plastic box you will find useful for serious projects. The best construction approach is to mount all of your parts on the top panel and leave the dish intact. There are different kinds of top panels available: aluminum sheet, perforated circuit board, and black plastic. The black plastic kind comes covered with a protective sheet of adhesive backed paper. Try not to remove this until you are completely finished with your drilling and cutting. The surface under the paper is very smooth—keep it protected as long as possible.

LAYING OUT A CONTROL PANEL

No matter what type of enclosure or panel you have selected, you are interested in making the best arrangement of controls and indicators on it that you can. Let yourself be guided by: convenience, feedback, and appearance.

Convenience and feedback are two important ideas from the science of **human factors.** The meaning of **convenience** is clear enough: Controls should be easy to manipulate and meters and other indicators should be placed where they are easy to see and well lighted. **Feedback** means that the way a control moves or feels tells you something about what effect it is having.

Here are some basic rules of human factors design:

1. Knobs for primary controls should be larger than knobs for secondary controls. Primary controls are those that you will be using most of the time; for example, the tuning controls on a radio receiver. Secondary controls are those you do not use as often, for example, gain controls and function switches.

2. If you are right-handed, group primary controls near the right edge of the panel; if you are left-handed, near the left edge.

3. Scale divisions on dials and meters should not be over-numbered or over-divided. On a scale from zero to 10, there should be major divisions for each unit, and at the most, one division for each half unit. It is enough if you number every other major division. Human-factor studies have shown that too many numerals and divisions are confusing and make it harder to read a meter scale or dial quickly.

4. Try to leave enough space so that you will be able to differentiate between similar-size knobs on the basis of their positions. This is more reliable than reading the label over the knob each time you reach for it. If you can't spread the panel out sufficiently, use different-size knobs.

5. If possible, the movement of a control should correspond to the movement of the meter or indicator with which it is associated. In other words, if your tuning knob rotates clockwise, your tuning scale should rotate clockwise—rot counter clockwise. Similarly, clockwise rotation of a control should not cause a meter reading to decrease (unless you are tuning the meter for a dip).

6. If there is no outside indicator associated with a control, the control function should always increase in a clockwise direction and decrease in a counterclockwise direction.

7. Toggle switches controlling on-off functions should be wired so that up is on and down is off.

The third consideration for panel layouts is appearance. You should consider this only after you have taken the human factors into account, but there is no reason for appearance to conflict with human-factor design. The key to a neat appearance is dividing the panel symmetrically and arranging controls and switches in straight lines. Take your time, cover the entire panel with masking tape, and divide the panel into four quadrants. Then mark the position of each control so that it lines up with the other controls and is neither bunched too tightly with other controls or placed unnecessarily close to an edge. Use the entire panel.

After you have used human-factor and appearance considerations to locate all of the controls, lights, plugs, meters, and dials on the masking tape covering the panel, check to make sure that there is no interference between adjacent pieces of hardware or between controls and the edges of the enclosure. When you are sure that there is no interference, you are ready to begin drilling and cutting.

LEGIBLE LABELS

Many home builders avoid putting labels on their projects entirely. They seem to feel that since they put the equipment together, they don't need any labels to tell them what the various controls and switches do. Other builders are content to label everything with embossed plastic tape. This at least tells other people who might be interested what a particular piece of gear is doing, but embossed tape gives a project an

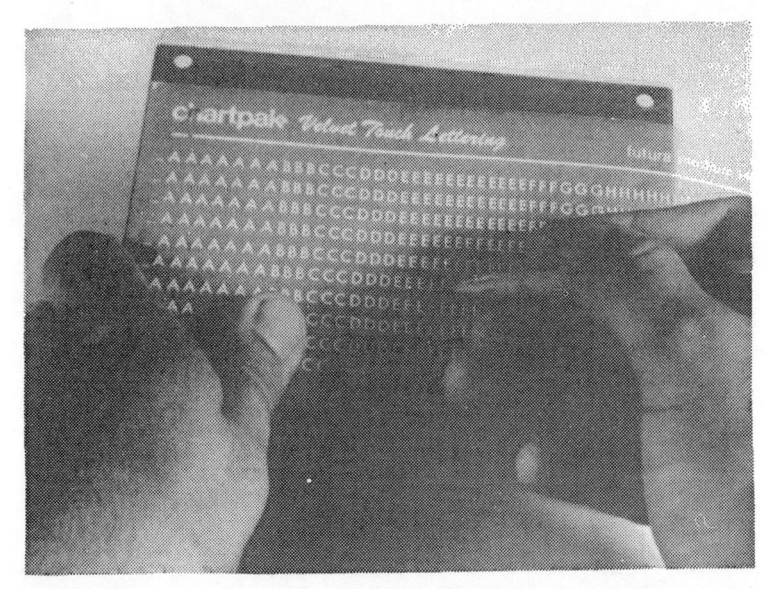

Fig. 3-5. Apply transfer letters by rubbing the front surface of the transfer letter sheet with a pencil or burnishing tool. Get the letters positioned just right before burnishing.

amateurish appearance. It isn't necessary to settle for no labels or embossed tape. There is a simple, inexpensive process that will produce labels just as good as those that are silk-screened onto expensive commercial equipment.

This process employs **dry transfer** letters (Fig. 3-5) that are available from art supply shops and electronics retail stores and catalogs. The letters are printed with a special material on the backs of plastic sheets. When you place the sheet over a smooth surface and rub on the front side of the plastic with a blunt pencil point or a ballpoint pen, the letter transfers itself from the plastic sheet to the surface beneath. Only the letter itself is transferred, so the end result is as clean as if it has been silk-screened.

As noted, you can buy sheets of transfer letters either at art supply stores or from radio suppliers. The ones at radio suppliers are usually printed with common electronics labels spelled out right on the sheet. The ones at art supply stores just give you several repetitions of letters, numerals, and punctuation marks. It is a matter of preference, but the ones with just the alphabets seem to be more economical in the long run, and they offer you a selection of type styles.

If you are going to pick type styles, you will have three decisions to make: What color, type face, and point size should the letters be?

For color, stick to white letters for dark panels, black for light panels. Other colors are available, including gold, but be careful or you'll wind up with a garish monstrosity of a panel that would be more appropriate to a Las Vegas slot machine than to your project.

Deciding on a type face, you will face an interesting choice. Human-factor researchers have shown that type with **serifs** is easier to read than **sans serif** type. Serifs are the little strokes at the tops and bottoms of letters like the ones in this book. Common names of type faces with serifs are Roman, Bodoni, and Caslon. While type with serifs may be theoretically easier to read, type without serifs—**sans serif**—looks cleaner, more modern, and less cluttered on an instrument panel. Common sans serif type faces are Gothic, 20th Century, and Futura.

No matter which face you select, you will also have to decide on the width of the strokes in the letters. Books like this are printed in lightface type. However, you will need a bolder type face for your panels to make the letters stand out. Select either a medium or a boldface type.

The **point size** tells you how big the type is. This book is printed in 9½-point type. Many other books use 8½-point type, which is just a bit smaller. On your projects, do not use anything smaller than 10 points or it will cause difficulty in reading the labels. You shouldn't have to use as much concentration to read a panel as you do to read a book.

All of the projects shown in this book use a typeface called **Futura**, in a medium width and a 14-point size.

The procedure for applying the letters is simple. Just place the plastic sheet in position, burnish the letter with pencil or pen point, and peel away the sheet. Here are some additional guidelines to help you achieve commercial-looking results.

1. For single words or groups of letters, select the point on the panel that corresponds to the center of the word. For instance, you may want the word "Volume" right under your

volume control pot. The middle of the word should be right under the middle of the control shaft. Start at this point, put the middle letter of the word, "l" in this case, right there, and work from the middle towards both ends.

2. Do not use all capitals. Combining initial capitals with lowercase letters in the body of the word gives a more pleasing appearance. For a modern, understated look, use lowercase letters for initial letters too.

3. You will save yourself a lot of headaches if you cut the sheets into sections of the alphabet. Sometimes it is even easier if you cut the sheets into separate lines. Cut-apart sheets are easier to store, also, if you put all the parts into clearly marked envelopes. You can mark the envelopes with transfers from title portions of the sheets.

4. Take as much time as you need to make your letters line up. Start your labeling in a position where you can use an edge of your panel as a guide for each letter. Continue using edges as guides, or use a label already applied to guide you.

5. Sometimes, especially on rough surfaces, letters do not transfer well. To overcome this difficulty, you can use the backing paper that comes with the transfer sheet to "prerelease" the letters. The backing sheet has a surface treatment that prevents it from picking up any transfer letters. If you put the backing sheet down on the table and the transfer sheet over that, you can rub on the letter without transferring it. This loosens the letter and helps it transfer easily to the rough surface.

6. After you transfer each letter, cover it with the backing sheet or with another piece of paper and rub over it with pencil or ballpoint to burnish it firmly in place. This may not appear necessary with every letter, but you can be sure that as soon as you think that you are doing just fine without this step, you will have a sudden accident and have to redo a whole word.

7. Sometimes, particularly on larger letters, the whole letter will not transfer the first time you rub over it. Be careful, then, when you peel the transfer sheet away from the letter that you hold it firmly in position until you are sure you have a whole letter. If you must burnish again, it is easier to reposition the sheet exactly if you have held it in place.

8. To protect your transfer letters, use a spray to cover them with a light coating of clear lacquer. A few words of caution, however: Be sure your spelling is correct **before** you spray, and do not spray clear plastic boxes unless you want to render them opaque.

SELECTING KNOBS, SWITCHES, AND INDICATORS

There are a great many different kinds of instrument knobs available today. Your selection will be based mainly on what appeals to your esthetic sense and what is readily available.

There are a few general observations to be made about knobs. First, do not mix several different styles on one panel. Your project will look a lot more professional if you use several sizes of the same style of knob. Second, make sure you have the right tool if you are using setscrew-type knobs. There is nothing more frustrating than finishing a project and discovering you have hex-head setscrews in your knobs and no Allen wrenches with which to drive them, or that your setscrew driver is a hair too wide to fit into the setscrew hole in the knob. Finally, if you have a rotary switch and a setscrew knob, you may find it advantageous to file a flat on the control shaft to prevent the knob from slipping.

A look through the pages of an electronics catalog will show you the kinds of knobs available. Besides ordinary knobs, you find that there are several kinds of **vernier** knobs that offer many turns of the knob to a single turn of the shaft. The cheapest vernier dials are planetary types with a fixed pointer and a rotating scale reading from zero to 10. Another type, which is not widely available, is a type with a moving plastic pointer and several concentric scales on which you can enter your own numbers. A third type looks externally like an ordinary knob. Inside, however, is a complete harmonic drive mechanism that gives a 40:1 tuning ratio with no backlash. One other type of knob is designed for use with multiturn potentiometers. This may appear to be a vernier knob, but is actually a **turns-counting** knob. The entire face is divided into 100 divisions over a full 360 degrees, and each time the indication goes through zero, a number in a window above the dial changes, indicating the number of turns from zero. These combinations of high-resolution multiturn potentiometers and turns-counting knobs are used primarily in analog computers.

As in the case of knobs, your choice of a switch will depend on your own personal preference and what is readily available. Generally, you will want to avoid the **slide** switch, even though it is the cheapest, because it requires a rectangular hole, which is hard to cut neatly. Slide switches also have generally poor reliability. **Rotary** switches are called for when there are a lot of poles to be switched or there are to be more than two positions. There are two kinds of rotary switches, the **shorting** type, also called **make-before-break,** and the **nonshorting** type, or **break-before-make.** If you have an application in which you cannot have the input to the switch open-circuited—for example, if you are switching speakers driven by power transistors—you must use the shorting switch. If you have a single load and several sources that you want to keep isolated, use the nonshorting type.

There is an almost endless variety of pushbutton and toggle switches, with many kinds of handles and lights that come on, go out, change color—or just sit there dumbly when you actuate them. Most of them are rated for several hundred volts and at least 3A. A few, notably the telephone type, are not designed for inductive loads and may burn away their contacts when connected to such loads. The only caution to observe with any of these switches is: Unless your project is very heavy, do not use pushbutton switches on vertical panels, or you may find yourself pushing a unit off the table when you only want to actuate a switch!

From a human-factors standpoint, a go/no-go "idiot light" indicator is more desirable than a meter or other indicator that has to be interpreted. In many cases, however, we want to know about trends that may be developing or whether a particular control is having an increasing or decreasing effect on some circuit parameter. In these cases, a meter is essential. The most common meters are round or rectangular, with a pointer that describes an arc along the meter scale. It has been shown, however, that these are not as easily read as edgewise meters in which the needle moves linearly along a long, narrow scale. The most effective installation for edgewise meters is vertical, so that the pointer needle moves up and down like the mercury in a thermometer.

Numerical readouts have been around for a long time in the form of neon glow tubes with stacks of numeral-shaped anodes inside. These have never been popular with home builders, however, probably because of their cost and the complexity of the logic required to drive them. New advances in light-emitting diodes (LEDs) have produced single-plane numerical displays that have minimal drive current requirements for high brightness and reliability.

At the same time, integrated circuit logic devices are becoming available that reduce the complexity of a complete decoder-driver to a single 16-pin dual in-line package (DIP). At the time of this writing, inexpensive solid-state 7-segment displays and decoder-driver packages are just becoming available to the home experimenter at reasonable prices. There is no doubt that this trend will continue, and that in a few years, use of these devices in home projects will be widespread. The Randomizer project in this book shows how these devices must be interconnected, what the various inputs are for, and what power and drive requirements are necessary.

As these components become readily available, there will be quite a rush to use them in all sorts of projects. While it is true that in terms of reading out a single number nothing beats a numerical display, it is also true that numerical displays are not as good as meters in showing whether a value is increasing or decreasing.

MAKING YOUR OWN METER SCALES

If all you are asking of a meter is to measure some common quantity like volts or amperes, you will find plenty of meters on store shelves to meet your requirements. You will even find a number of meters calibrated in volume units (VU) or decibels. However, beyond these choices, there is little selection. For example, if you want a meter that measures transistor beta and leakage (Fig. 3-6), as in the BJT-FET transistor checker project in this book, you will not find a single meter on the market that is so calibrated.

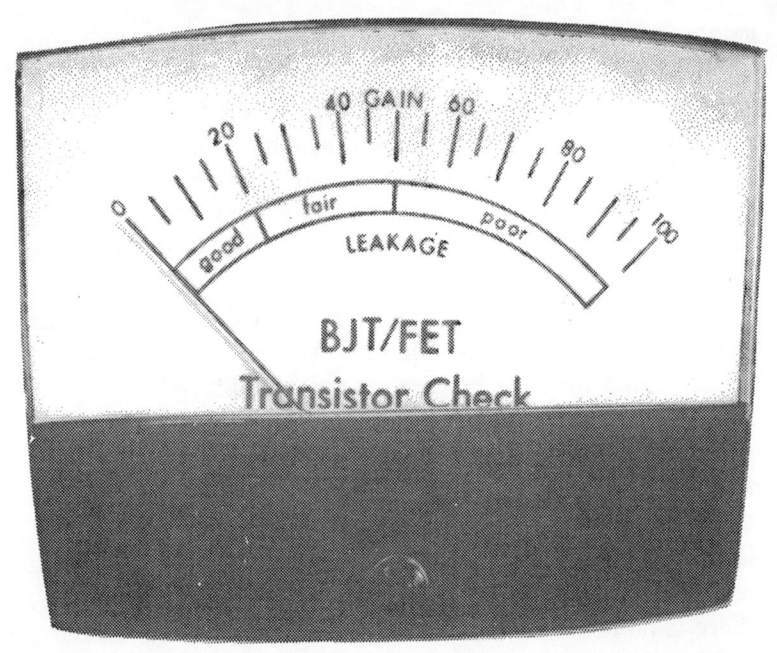

Fig. 3-6. An ordinary one-milliampere meter was modified as explained on these pages to produce this unique meter.

It is neither expensive nor especially difficult to prepare custom meter scales. The main ingredients are patience, some drawing tools, and a painstaking approach. The steps are outlined below.

1. Select an appropriate meter. You will want one that spreads your scale out over the entire range and in which the cover can be separated from the works. On most old-fashioned round-faced meters, there are three or four screws on the back that can be loosened to allow the backplate and meter works to slide out of the housing. On the common, inexpensive square-faced meters, the faceplate has been snapped onto the back by means of four raised tabs on the back surfaces of the faceplate that mate with recesses on the rear shell. If you are very patient, you can shave these tabs down with a very sharp X-acto knife, allowing you to remove the faceplate. Later on, you will have to reassemble the faceplate to the rest of the meter and use a drop of cement in place of each of these tabs.

2. Take out the two tiny screws holding down the old meter scale, put them where you will be able to find them again, and

84

remove the old scale. Be very careful at this stage not to damage the meter needle or the very fine bearings it rides on. By this point you have voided all warranties

3. Put the meter aside and tape the scale down towards the bottom of an 8½x11 in. piece of paper. Very carefully line a ruler up with the lines corresponding to the maximum and minimum scale divisions and project these lines inward. The point where these two lines cross is theoretically the pivot point of the meter needle.

4. You should have already decided what scale divisions you want to use: all of them, every other one, every third one, etc. Extend a line from the pivot point on your paper through every scale division you intend to use. Extend the lines a fair distance, at least twice as far as the distance from the pivot point to the outer end of the division mark on the meter scale.

5. Use a circle template or a drafting compass to measure the radial distance from the pivot point to outermost point of the division marks on the original meter scale.

6. Double this distance and strike off an arc that intersects your projected division lines. This gives you the basis for a meter scale that is exactly twice as large as the one that was originally in the meter.

7. Remove the original meter scale and take a heavy pencil and darken each of the scale graduations, starting at the arc that you drew with the compass and carrying them inward toward the pivot point, ⅜ in. for major divisions and ¼ in. for minor divisions.

8. Place a clean piece of paper over the paper you have been using, and carefully and neatly trace the division marks you have made. Ideally, you should use a drafting pen, but you can get satisfactory results if you use a sharp medium-soft lead pencil and are careful not to smudge. Avoid ballpoints and felt tip pens. If you use a pencil, do not make any erasures. If you "goof," start over on a new piece of paper. When you finish, spray the sheet with a workable fixative such as Krylon, which can be purchased at an art or engineering supply store.

9. Apply transfer letters as described earlier in this chapter to identify the scale divisions and to add any other

information that you want on the meter face. Since your final product will be half as big, you may want to use a larger point size for some of the lettering. Generally, however, 14-point lettering, reduced to 7-point in the final size, will be just right for your meter.

10. This is a step you can't do yourself, but it can be done cheaply and easily for about $2. Check your Yellow Pages under "blueprinting" or ask a local printer for the name of a nearby laboratory that has a process camera, and ask for a 50-percent-reduction positive print of your meter scale. This is a routine job for this kind of business, but be sure they understand that the reduction must be **exactly** 50 percent. Most of the time you should be able to get overnight or same-day service.

11. Take your new scale home and, using a strong light, align it until it exactly matches the division marks on the original meter scale. Hold or tape the two scales together and carefully trim the outside of the new scale to exactly match the contour of the old one.

12. Use rubber cement or a spray-on contact adhesive to attach the new scale firmly to the old. Then remount the scale on the meter and reinstall the faceplate.

RANDOMIZER PROJECT

Now that 7-segment LED displays and IC decoder-drivers are becoming available at reasonable prices, the home experimenter will want to know how to incorporate these devices in his projects.

While the one-digit Randomizer (Fig. 3-7) isn't the most useful piece of equipment in this book, it was selected because it is an ideal project for demonstrating 7-segment LED techniques. There is a minimum of extraneous circuitry to cloud the issue. Of course, the Randomizer is useful as an adjunct to games in which dice are normally used, and some may even find it useful for studying probability. It is a good showoff project too, since the clear plastic box reveals the simplicity of the Randomizer's inner workings.

You may find that some of the parts used in the Randomizer are not readily available where you live. However, other similar parts are available, most notably from Radio Shack, and you should have no trouble modifying the layout to make the design fit the parts available.

LED Display

The most notable feature of the Randomizer is the 7-segment LED display. This is a big one; the digit on the display measures a full 0.625 in. high. The viewing distance for this size display is a full 17 ft in normal room light. Since the display achieves this at only 5V and 177 mA with all segments illuminated, you can see why these displays have become so popular. The light source for each segment is a semiconductor diode that emits light when current flows through it.

Decoder-Driver

You can make the diodes light up any time you want to in any combination you wish merely by arranging for 25 mA to flow through each diode, but in practical cases, you will want to drive the display with digital logic. The display used in the Randomizer incorporates an integral bcd (binary-coded decimal) to 7-segment decoder-driver and current-limiting resistors on a printed circuit board permanently attached to the display module. Integrated circuit decoder-drivers are also available separately at surprisingly modest cost.

Fig. 3-7. The Randomizer. Display changes each time CYCLE button is pushed.

Ring Counter

The decoder-driver is driven by another integrated circuit, a readily available divide-by-10 circuit connected as a ring counter. The ring counter is, in turn, driven by a simple clock that employs a unijunction transistor (UJT) pulse generator. The clock frequency is 1000 Hz. This means that 100 times every second the ring counter counts from zero to 10.

The clock is turned on and off by a momentary-contact pushbutton switch. Since the human reaction time is very much slower than the clock-cycling rate, the effect is that many, many 0-10 cycles occur during the shortest time the pushbutton is depressed, and the number that appears on the display when the clock stops is a true random digit.

Construction

The Randomizer was constructed inside a 2x2x1¼ in. clear plastic box for the sake of novelty. To preserve the impression of compactness, the batteries are kept separate, and power is applied through a miniature phone jack. The display and decoder-driver are already interconnected in a single unit. If you are using separate displays and decoder-drivers, Fig. 3-8 shows how to connect the output of the decoder-driver to the display. Some displays include built-in limiting resistors; others do not. Be sure you know which kind you have. If you apply full voltage with no limiting resistance to your display, you stand a good chance of destroying both the display and the driver.

The UJT clock and IC ring counter are mounted on a separate printed circuit board (Fig. 3-9), which was planned so that short wire jumpers could be run directly to connect it to inputs on the display board. The jumpers are stiff and hold the two boards in the proper relative positions. The box has a hole cut in it for the face of the display, and the display is held in place by a few sparingly applied drops of cement.

Randomizer Circuit

The clock is a conventional UJT relaxation oscillator. You will note from Fig. 3-10 that the clock input to the ring counter is designated "$\overline{C}$." The bar over the C indicates that the clock

Fig. 3-8. Schematic for Randomizer.

C1 0.1 uF, 12V dc disc ceramic
C2 0.002 uF, 12V dc disc ceramic
IC1 HEP C3800P, IC decade counter (installed in 14-pin socket)
IC2 Dialight 730-007 LED 7-segment display with integral Texas Instruments SN7447N BCD to 7-segment decoder-driver and 180-ohm resistors.
Q1 Archer (Radio Shack) 276-111 unijunction transistor
R1 10K, ¼W
R2 470 ohms, ¼W
R3 220K, ¼W
S1 Momentary-contact pushbutton switch, spst

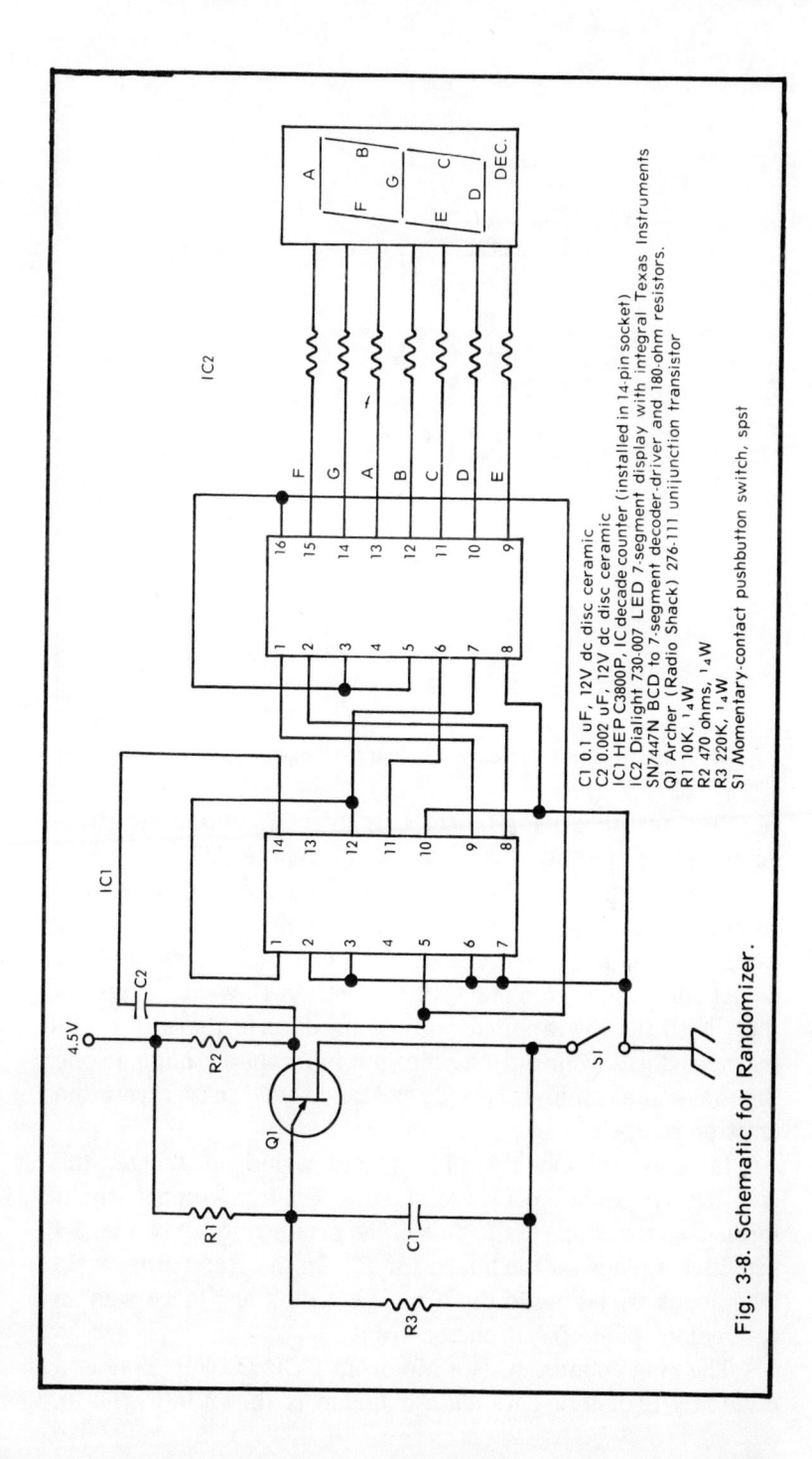

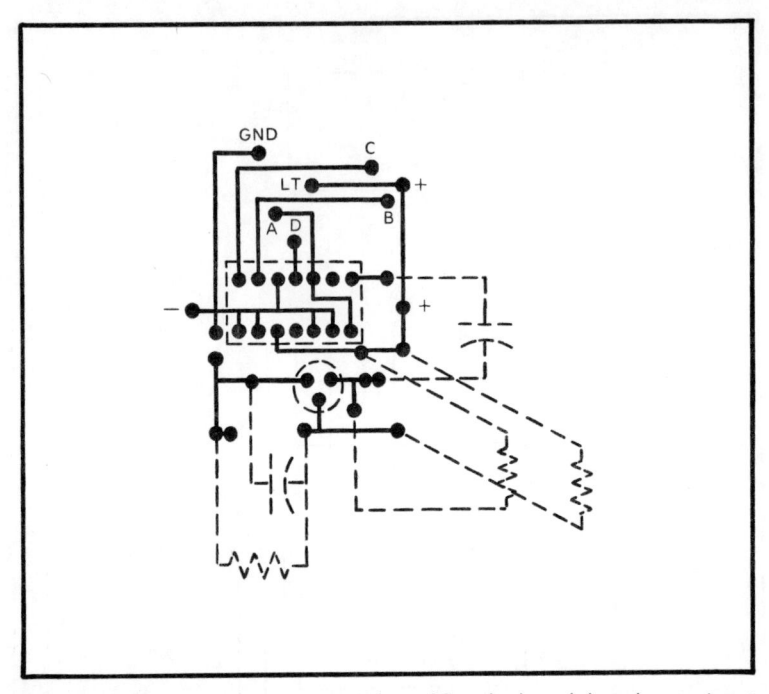

Fig. 3-9. Actual-size printed circuit board for clock and decade counter of Randomizer.

requires negative-going pulses to trigger. To obtain negative-going pulses, the output is taken from B2 of the UJT.

The inexpensive UJTs used for this project presented something of a problem. The first couple that were tried wouldn't oscillate. Theoretically, the characteristics of UJTs do not vary much from one type to another. However, it seems that, with the low applied voltage used here, there is a difference. Check your unijunctions out beforehand, and use one that has a maximum forward resistance of 300 ohms from the emitter to each base.

Initially, the clock would run but would not trigger the counter. The addition of R3 solved this problem. Apparently, it influences the shape of the pulse. The printed circuit of Fig. 3-9 provides a place on the board for R3. In the Randomizer, R3 was "back wired" onto the circuit board, since there was no convenient place for it on the front.

The ring counter IC is a Motorola HEP C3800P. This is a divide-by-10 counter; its logic diagram is shown in Fig. 3-10.

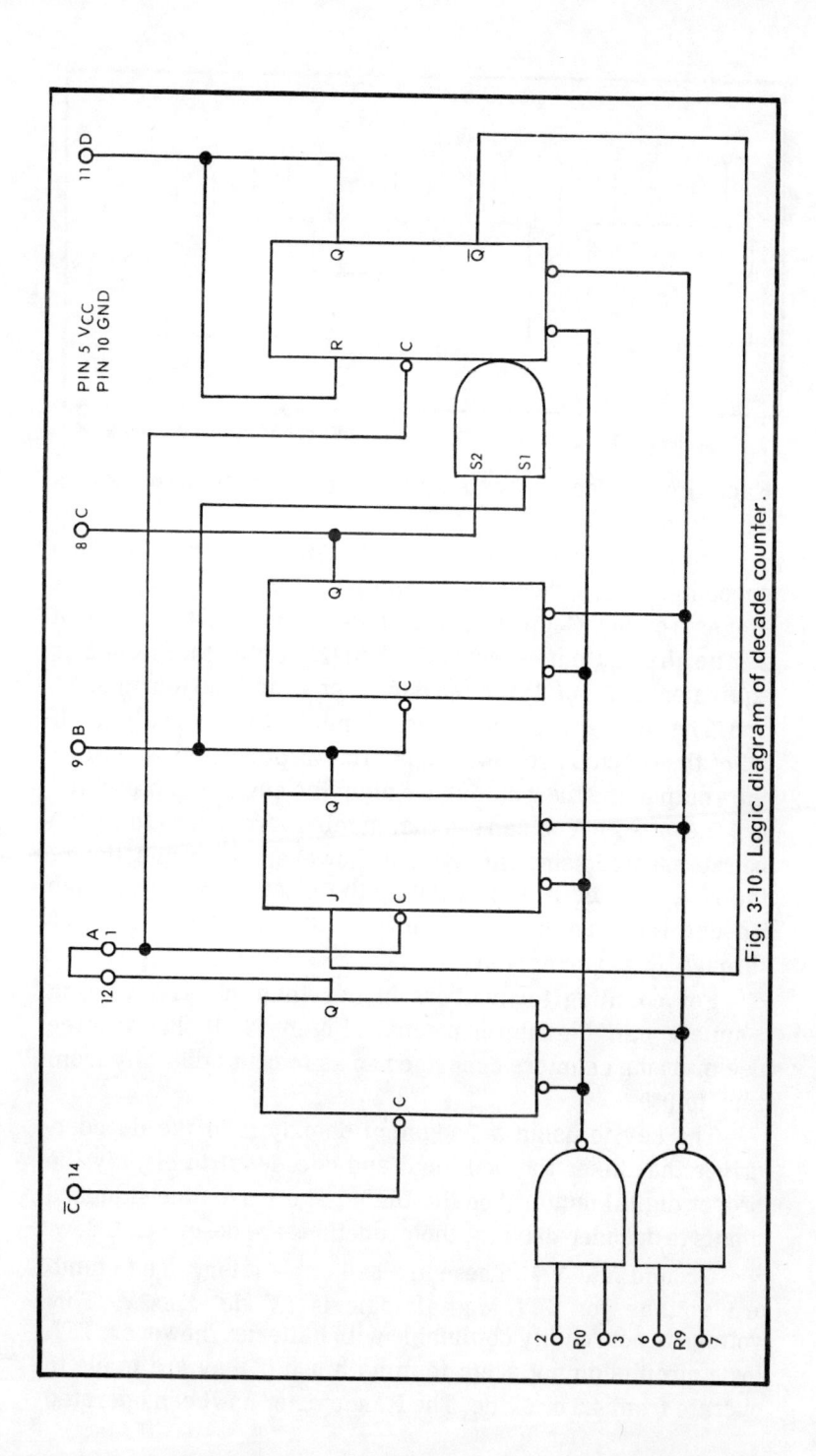

Fig. 3-10. Logic diagram of decade counter.

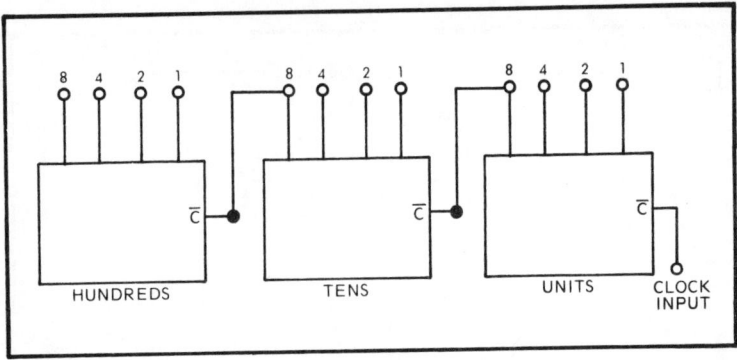

Fig. 3-11. Decade counters cascaded to permit counting from zero to 999.

Each time a clock pulse is applied, the counter increases its output by one.

The output format is binary-coded decimal, or bcd. In this encoding system, there are four outputs that can be either **high** (**logic 1**) or **low** (**logic 0**). The outputs correspond to powers of 2. The first output is equivalent to 2^0, or 1; the second is equivalent to 2^1, or 2; the third to 2^2, or 4; and the fourth to 2^3, or 8. The number represented is the sum of the high outputs. If all of the outputs are low, the number represented is zero. If the 4 output and the 2 output are high, the number represented is 6 (i.e. 4 + 2). You can see that numbers from zero to 15 can be represented using this system; however, the connection of the ring counter is such that on the clock pulse after each 9 count, the counter is reset to zero, so the numbers from 10 through 15 never appear.

For counting to numbers higher than 10, several ring counters can be interconnected. Figure 3-11 shows three decimal ring counters connected so as to count digitally from zero to 999.

The key to using a 7-segment display is in the decoder-driver that takes the bcd input and decodes it to display the proper digital numeral on the LEDs. There are nine standard inputs to decoder-drivers; their functions are described below.

Ground and V$_{cc}$. These are self-explanatory. The standard voltage for TTL digital logic is 5V dc ±0.25V. This voltage is not readily obtainable with batteries, however, TTL logic circuits do not seem to mind much if they are made to operate from 4.5 or 6V dc. The Randomizer has been operated

92

with both values, and the effect on the display brightness is not discernible. However, with 6V applied, after a certain number of cycles the counter would appear to "hang up." This does not happen with 4.5V, although as the batteries get older and voltage drops, the oscillator may stop working until the batteries are replaced.

A, B, C, and D. These correspond, respectively, to the 1, 2, 4, and 8 outputs of the ring counter. In some cases, the ring counter designates these outputs as **Q0, Q1, Q2,** and **Q3**.

LT (Lamp Test). This input **must** be high for the display to show the count. If it is low, all seven segments will light up, regardless of what the other inputs are.

RBI (Ripple-Blanking Input). This input has an effect **only** if the bcd input is zero. In this case, if the rbi is high, the zero will be displayed. If it is low, the zero will be blanked. This is used with the **ripple blank** output of adjacent displays to suppress leading and trailing zeros. (See Fig. 3-12.)

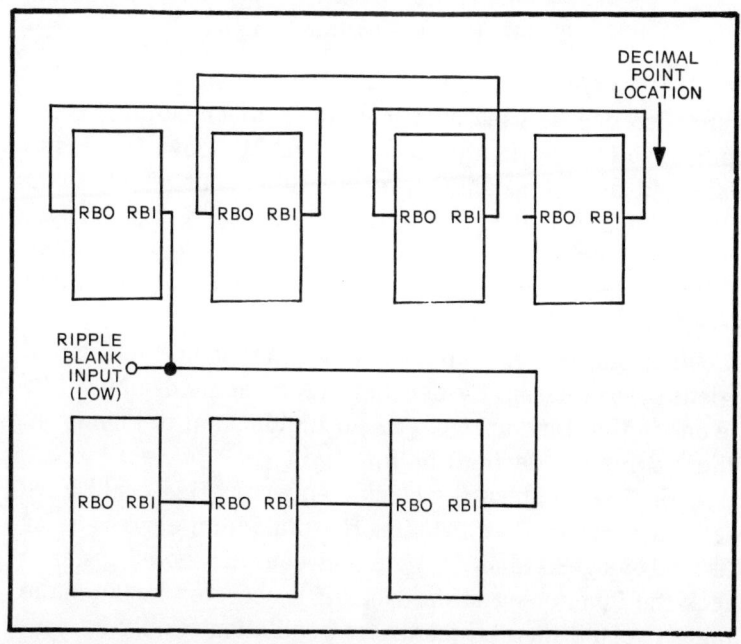

Fig. 3-12. Decoder-driver ripple blank input and output connections are used to blank leading and trailing zeros in multidigit displays. If the input to the seven decoder-drivers shown were 2.03, displays connected to the decoder-drivers would show "2.03" if ripple blank input were low, but "0002.030" if rbi were high.

Fig. 3-13. FET-set shortwave receiver.

BI-RBO (Blanking Input—Ripple-Blanking Output). This connection can be used as either an input or output. As an input, it will blank the display whenever it is low. If it is not connected, as in the Randomizer, it will have no effect. As an output, it will be high except when the display is blanked by a low rbi, when it will be low. (Again, see Fig. 3-12 to understand how rbi and rbo work together to suppress leading and trailing zeros.)

On displays in which there is a decimal point, the decimal logic is separate from the decoder-driver. Some displays, like the one in the Randomizer, ground the decimal to turn it on. Others drive the decimal high to light it.

A final word about the clock frequency. This could have been much higher than 1000 Hz. The maximum clock rate for this kind of logic is about 18 MHz. However, it seemed easier to check the functioning of the clock if it operated in the audio range. It could then be tested for oscillation simply by connecting a high-impedance headphone across the UJT output. And the very low-speed cycling of the display has a modulating effect, so that the 8 that is displayed when the clock is running is noticeably less bright than numerals displayed when the clock is stopped. This modulation of the light output is further indication that the circuit is operating properly.

FET-SET SHORTWAVE RECEIVER PROJECT

With this project, illustrated in Fig. 3-13, you can listen to distant shortwave stations in the 3.5-8 MHz shortwave band. The signals you will hear will include hams on the 80- and 40-meter amateur bands, domestic and overseas shortwave broadcast stations, and the 5 MHz time broadcasts of U.S. Government station WWV. The circuit takes a very old detector idea and gives it a new twist with a dual-gate MOSFET. The detector part of the circuit is actually a minor modification of a circuit that appears in the "RCA Transistor Manual." The FET-Set shortwave receiver gives you the option of listening on its own built-in speaker or on a private earphone.

Shortwave Receiver Circuit

The signal from the antenna is applied through C1 to the tuned circuit consisting of L1, C2, and C3. For the sake of stability, it is very important that C1 connect to the coil at the tap point, one-quarter of the way up from the ground end of the coil. Capacitor C2 is the main tuning control, and C3 provides bandspread capability. Under the heading of "Construction," we will explain how to make the bandspread capacitor by removing plates from a larger unit. By a happy stroke of luck, the modified bandspread capacitor provides almost exactly 10:1 resolution relative to the main tuning capacitor.

The regenerative detector is an old concept in radio receivers—something quite simple, and at the same time, quite clever. Although it is—bar none—the most sensitive type of detector, it has some disadvantages that limit its use to inexpensive projects like this. Its major disadvantage is that it **blocks** (loses its sensitivity) in the presence of large signals. Also, its response is less than perfectly linear (it distorts). Given these limitations, the regenerative receiver is still an excellent way to squeeze a lot of sensitivity out of a very few components.

How does the regenerative detector work? The basic concept involves an amplifier in which a little bit of the output is fed back, in phase, to the input. Think of the circuit as starting with just the gain available from the amplifier without feedback. If a little of the output is fed back into the input, this effectively makes the input signal a little bigger than it would be all by itself. In turn, this makes the output a little greater, and this boosts the input just a little more. In effect, the signal picks itself up by its own bootstraps. If the amount of signal that is fed back to the input is too great, the amplifier will go into self-oscillation and no signal amplification will take place. The amount of feedback, or **regeneration**, must be controlled so that the circuit operation is just below the point of self-oscillation.

The usual method of obtaining regeneration is by means of a separate winding, called a **tickler**, on the tuning coil. The circuit in our FET-Set uses a different approach. (Refer to Fig. 3-14.) Field-effect transistor Q1 is a dual-gate MOSFET; that is, it has two control elements. This kind of FET was developed for high-frequency applications, where interconnections can be made that minimize the device's apparent capacitance. For our circuit, though, we will use gate 1 for the signal input and gate 2 for a regeneration signal coupled back from the FET drain.

Even with all of the amplification available from the regenerative detector stage, there is insufficient audio at the drain of Q1 to drive an audio amplifier stage. Consequently, we have included Q2, a FET preamplifier. While there is nothing remarkable about the RC circuit used to couple the output of Q1 to the input of Q2, we should explain why the output of Q2 is coupled to the input of the audio amplifier through a matching transformer. This is necessary because Q1 and Q2 are both n-channel devices and their ground bus is negative, while the audio amplifier uses a positive ground bus.

While this project was still in the design phase, I considered designing and building an audio amplifier custom tailored for this application. It turned out, however, that there are several audio amplifiers on the shelves in the radio hobby stores that are well suited for this project and cheaper to buy than the parts for a scratch-built amplifier. The kit audio

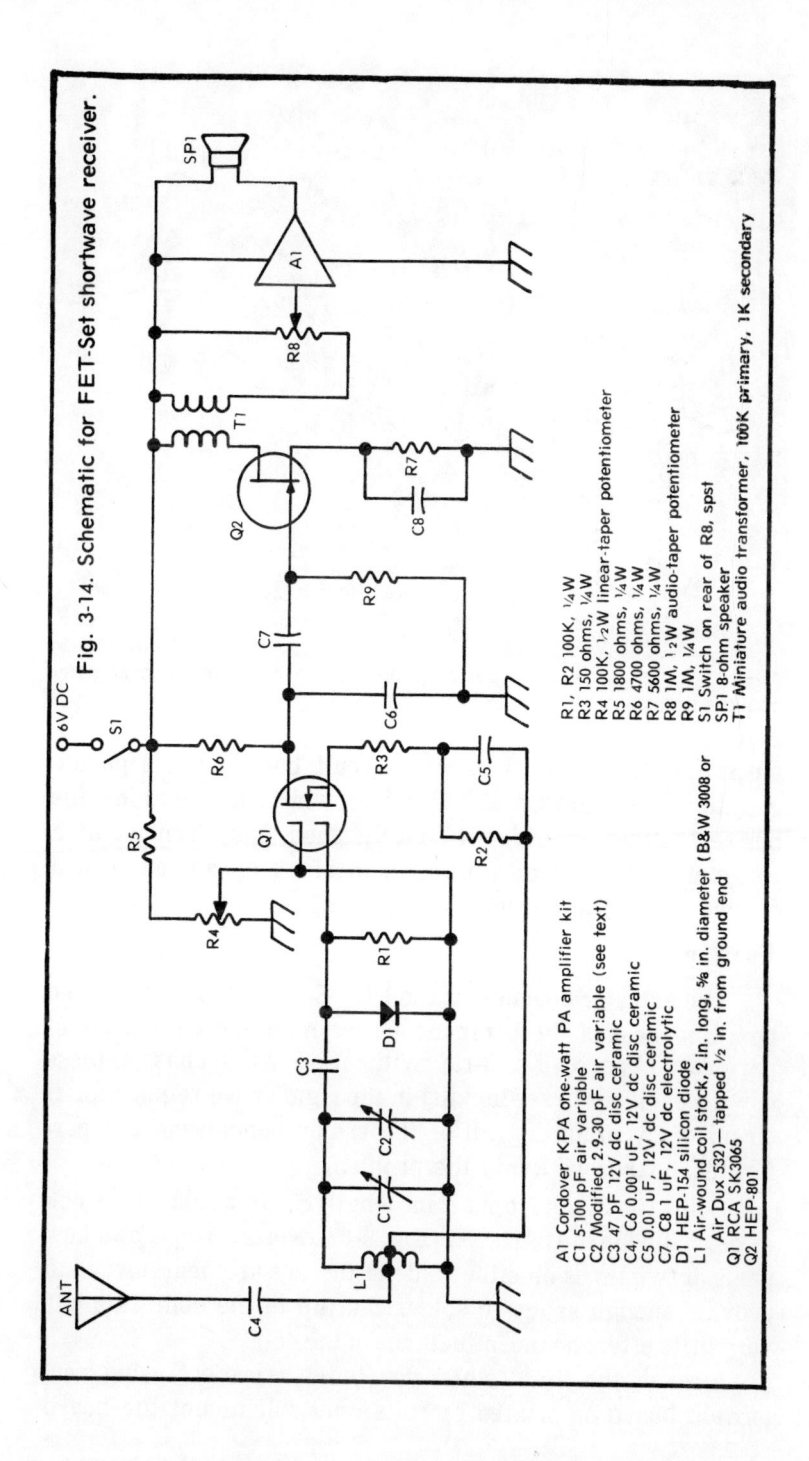

Fig. 3-14. Schematic for FET-Set shortwave receiver.

A1 Cordover KPA one-watt PA amplifier kit
C1 5-100 pF air variable
C2 Modified 2.9-30 pF air variable (see text)
C3 47 pF 12V dc disc ceramic
C4, C6 0.001 uF, 12V dc disc ceramic
C5 0.01 uF, 12V dc disc ceramic
C7, C8 1 uF, 12V dc electrolytic
D1 HEP-154 silicon diode
L1 Air-wound coil stock, 2 in. long, ⅝ in. diameter (B&W 3008 or Air Dux 532)—tapped ½ in. from ground end
Q1 RCA SK3065
Q2 HEP-801

R1, R2 100K, ¼W
R3 150 ohms, ¼W
R4 100K, ½W linear-taper potentiometer
R5 1800 ohms, ¼W
R6 4700 ohms, ¼W
R7 5600 ohms, ¼W
R8 1M, ½W audio-taper potentiometer
R9 1M, ¼W
S1 Switch on rear of R8, spst
SP1 8-ohm speaker
T1 Miniature audio transformer, 100K primary, 1K secondary

Fig. 3-15. Most of the receiver components are mounted on perforated board using push-in terminals.

amplifier consists of a printed circuit board with separate resistors, capacitors, and transistors, which the buyer installs himself. Other amplifiers, equally good, may consist of a complete assembly encapsulated in black epoxy, or even a one-watt audio amplifier IC.

Construction

The exterior housing for the FET-Set is a 5x5x5 in. cowled box, but most of the components are mounted on a piece of perforated board (Fig. 3-15). Without an inside chassis, these cowled enclosures are lacking in the rigidity we require for rf circuits; by mounting all of the rf components on the perforated board, we avoid the problem.

Coil L1 is made from a 2 in. length of air-wound coil stock as described in the parts list (Fig. 3-14). Figure 3-16 shows how parts of two turns on either side of the tap are bent inward to provide enough space to solder the tap to the coil. This has very little effect on the inductance of the coil.

Most of the time, when you build a project using perforated board or printed circuits, you will mount the board inside the enclosure with some kind of standoff brackets. In the case of the FET-Set, though, we have the opportunity to employ a unique method of mounting the board to the enclosure. The tuning knobs for the main tuning and bandspread capacitors are planetary vernier units that mount directly to the front panel of the enclosure. On the back of each knob is a ¼ in. (inside diameter) bushing with a setscrew. The shaft of the tuning capacitor fits into the bushing. If you are very careful to make the locations of the vernier knobs on the front panel correspond to the locations of the capacitors on the perforated board, you can use the shafts of the capacitors to support the circuit board (Fig. 3-17).

It isn't as hard to achieve this matching of holes as it sounds. The secret lies in the pattern of holes on the perforated board. If we select two holes on the board as the starting point for drilling the holes for the variable capacitor shaft, we can use the same two holes to mark the front panel of the enclosure with great precision. In fact, in this case, since the regeneration control is mounted on the perforated board, we

Fig. 3-16. By bending adjacent turns of the coil inward as shown, it becomes possible to solder a tap to the proper turn without interference.

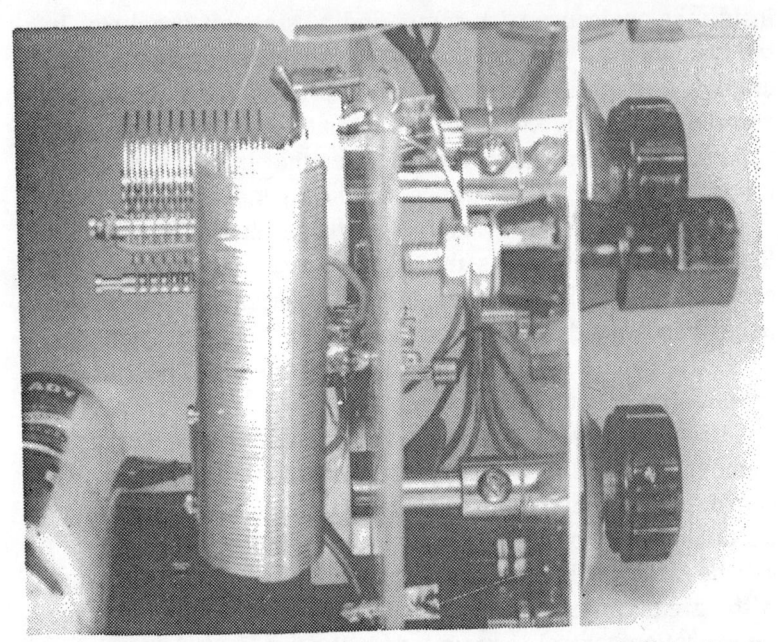

Fig. 3-17. The tuning capacitors are screwed to the perforated board, and the shafts of the capacitors themselves are used to attach the assembly to the front panel.

want to mark its location on the front panel in the same way. If we take our time and make sure that our drill bit lines up exactly with the marks on the front panel, we can be sure that the holes in the panel and circuit board will match.

The main tuning capacitor is an unmodified 100 pF air variable. The bandspread capacitor requires some modification, however. Starting with a 30 pF unit similar to the 100 pF main tuning capacitor, but with fewer plates, we proceed to remove more plates. When we are finished, we want to have only two stator plates and two rotor plates (Fig. 3-18). To do this, we use long-nose pliers and our bare hands to bend each plate, in turn, out away from the body of the capacitor, where we bend it back and forth until it breaks away from its mounting points. Be very careful not to damage the plates that remain on the capacitor. The last thing to do when only the three plates are left is to check to make sure that they are all parallel and do not short together at any point in their rotation.

The battery holder for the receiver is mounted on the bottom plate of the enclosure. For batteries, C-cells were used because the audio amplifier requires appreciable current, and smaller batteries would not be able to supply this current for long periods of listening. The speaker, earphone jack, and audio amplifier all are mounted on the back panel of the enclosure. Figure 3-19 demonstrates how conducive neat cable lacing is to a neat appearance.

FET-Set Operation

The key to good shortwave reception with this or any receiver is a good antenna. For these bands, you'll want a long, high wire—the longer and higher, the better. How long is long? Consider 50 ft an absolute minimum for reliable reception. You will also need a good ground for your antenna to work with. A "good ground" means a solid connection to a water pipe or copper rod, or pipe driven at least 5 ft into damp soil. A barely adequate ground might be a screw attached to an electrical junction box in the room.

Having said something about good antennas and grounds, one must recognize that people have used all sorts of bad or

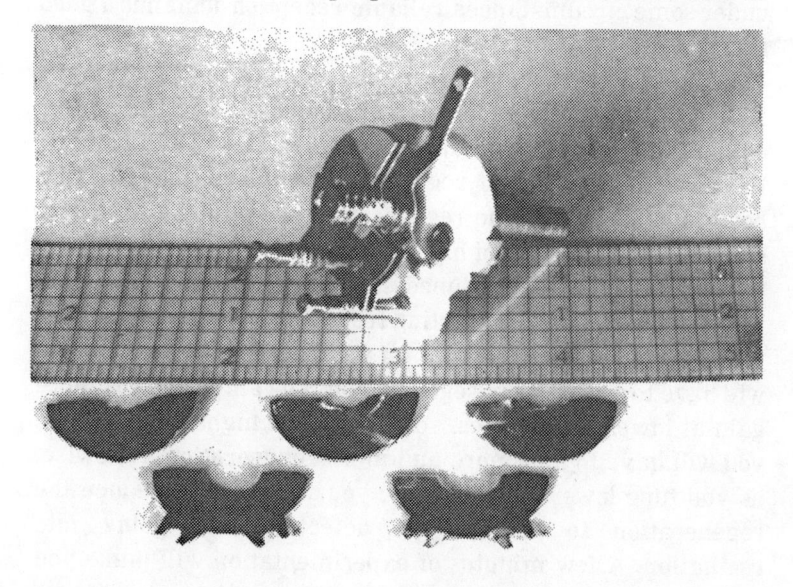

Fig. 3-18. By removing plates from a standard variable capacitor, we are left wqth a unique capacitor of very low capacitance.

Fig. 3-19. Interior view of FET-Set.

mediocre antennas and grounds with surprising results. I have logged Radio Japan using the FET-Set with an antenna made from a short length of wire clipped to the finger stop on a dial telephone! The important thing to understand is that while bed springs and window screens may produce adequate results under some circumstances, **reliable** reception demands a good antenna and ground.

Once you have your antenna and ground connected to the FET-Set, turn the regeneration all the way down, the power on, and the volume up to about three-quarters of full volume. Increase the amount of regeneration until you hear a loud howl, and then back the regeneration off until the howl just stops. This is the point of highest sensitivity for the frequency to which the receiver is tuned. You should be able to tune a little ways around your starting frequency and hear a number of signals. If you tune very far off your starting frequency, you will have to readjust the regeneration. The amplifier has more gain at lower frequencies, so as you tune higher and higher, you will have to add more and more regeneration. Likewise, as you tune lower in frequency, you will have to reduce the regeneration to prevent the detector from going into oscillation. A few minutes of experimentation will make you an expert at keeping the sensitivity of the receiver at optimum without slipping into oscillation.

What Will You Hear?

At the low end of the tuning range, you can expect to hear ham signals in voice and Morse code. As you increase the frequency, you will encounter some warbling signals characteristic of Teletype, some droning signals that sound like airplane engines and are facsimile transmissions, and some commercial stations broadcasting Morse code.

At right about 5 on the vernier dial, you will encounter the regular tone and voice announcements of WWV on 5 MHz. (WWV broadcasts the same signal on 2.5, 10, 15, 20, and 25 MHz from Boulder, Colorado. The National Bureau of Standards also operates WWVH in Kauai, Hawaii. WWV uses a male voice to make a time announcement each minute, and WWVH uses a female voice a few seconds earlier. In Portland, Oregon, the FET-Set picks up both signals.) If you listen regularly to WWV, you will not only have the most accurate wristwatch on the block, but you will also be advised of significant meteorological events during the 19th minute of each hour and of the projected quality of radio propagation across the Atlantic Ocean for the next 24 hours. This information is broadcast during the 15th minute of each hour. The broadcasts from WWV and WWVH do not make the most compelling listening on the shortwave band, but they do contain a wealth of information. You can learn more about this information from "NBS Special Publication No. 236, NBS Frequency and Time Broadcast Services." You can get it by sending 25 cents to the Superintendent of Documents, U.S. Government Printing Office, Washington, D.C. 20402.

As you tune above WWV, you will encounter several signals from the Voice of America and a number of overseas broadcast stations. Sometimes these stations broadcast in their native languages; frequently they broadcast in English. Tuning higher you will encounter the 40-meter ham band.

You will find that reception on these bands is better in winter than in summer, and better at night than during the day. The reason for this is that in winter there are fewer thermal-type thunderstorms to create static. And at night and during the winter, conditions are more favorable in the ionosphere, which is responsible for the reflecting radio waves over long distances.

4
Troubleshooting Your Projects

Like everyone else, you eventually will have the experience of building a project and then finding that it doesn't work when you turn it on. The likelihood of this happening depends on the nature of your project. If you build a kit offered by a reliable manufacturer such as Heath or Eico, chances are pretty good that your project will work the first time you turn it on. If you are building a project based on an article in a popular magazine, your chance of immediate success is good to fair, depending on how closely you follow the author's recommendations and how well the author and magazine editors check out the article. Slipups are common enough that it is sometimes wise to delay starting an expensive or complicated project 2 or 3 months to see whether any corrections appear in the magazine. If you decide to put something together on your own, based on some circuit ideas you have seen, say, in a transistor handbook, don't be disappointed if you don't get the results you hoped for the first time you apply power.

But don't let this discourage you. Engineers expect a "debugging" period as a part of the design of any complicated piece of apparatus. Even with established designs that are being put together on assembly lines every day, enough failures occur to keep quality control inspectors on their toes.

When you find yourself with a project that doesn't work, you will find that it helps to go through a logical series of steps to isolate the problem or problems. (Don't assume that problems only come one at a time!) This chapter will help you to proceed sensibly, one step at a time, to locate and eliminate trouble spots.

TROUBLESHOOTING EQUIPMENT

You already own some very sensitive troubleshooting equipment that hasn't cost you a thing—your senses. Use your eyes to look closely at solder joints and components. Can you smell any burnt components? Hear that transformer sizzling? Its output must be shorted. Is that resistor hot to the touch? It's carrying too much current. Some people even claim to use their tongues to check low voltages, but I have better uses for my tongue and prefer to take these reports with a grain of salt.

Once you have exhausted the possibilities of your built-in test equipment, you will find that there are some items you will have to purchase. Some are absolutely essential, and some are nice to have around but indispensable only in certain limited cases. Many of them are described below.

Meters

The most common meter for troubleshooting is the **volt-ohm-milliammeter**, or **VOM**. As its name indicates, this is an instrument that is capable of measuring voltage (ac or dc), resistance, and current (dc only). Prices vary, and some very good imported models sell for less than $10.

VOMs differ in the internal resistance of their voltmeter sections. Very poor VOMs have a meter resistance of only 1000 ohms/volt (ohms per volt). Better units provide 20,000 or 30,000 ohms/volt, and expensive types may provide 50,000 ohms/volt. The number of volts referred to in the ohms/volt rating is the top voltage of the meter's range. In other words, when a 20,000 ohms/volt meter is set to its 0-100V range, its internal resistance is 20,000 times 100, or 2 megohms.

When we measure voltage drops across large resistances, the internal resistance of the meter is very important, since the meter can become part of the circuit and affect the voltage level. **Electronic voltmeters**, or **EVMs**, may be used instead of VOMs to obtain more accurate measurements. Early electronic voltmeters used vacuum tubes and were called **vacuum-tube voltmeters**, or **VTVMs**. Some electronic voltmeters now employ field-effect transistors—which, like vacuum tubes, have a high input impedance—in transistorized units similar to VTVMs. These are often referred to as **FETVMs**, for **field-effect-transistor** voltmeters.

VOMs, FETVMs, and VTVMs are all used to measure resistance in ohms. They do this by providing an internal voltage (which is applied to the unknown resistance) and measuring the current through the resistance in question. It is important that you do not try to measure a resistance in a circuit to which power is applied. There could be enough voltage across the resistance to permanently damage the meter.

While VOMs and FETVMs are both used to measure current, most VTVMs do not have a current scale. It is very important that you understand that voltages are measured with the meter in **parallel** with the part of the circuit in question, but that current is measured with the meter in **series** with the circuit. Here's why: The meter used for current measurement has some very small-value resistances, called **shunts**, inside it. Each shunt is made for a different current range and is selected by means of a switch on the front panel of the meter. The shunts are made of a few turns of very fine nichrome wire, and they are precision resistors of fractions of an ohm. The meter measures current by measuring the voltage drop caused by the current as it flows through one of these shunts. If you connect one of these shunts across any appreciable voltage or in a circuit in which more current than the shunt design capacity is flowing, the shunt will overheat and burn up.

There are some test meters available that use a **digital** display for their output. These are still in the $200-plus category, but decreasing costs of ICs and LED displays should bring their cost down to more attractive levels soon. You can already buy a complete 4-digit digital voltmeter IC with BCD output for $16 in lots of 100.

Signal Generator

You can do a lot of troubleshooting with just a VOM or EVM, but you can speed up your checking considerably if you have a signal generator. If you are testing audio circuits, you can get by with a single-tone audio generator, although high-fidelity testing may require a precision generator with a calibrated output so that you can observe the frequency response of an amplifier.

If the project you are troubleshooting has radio-frequency stages, you will want an rf generator. There are some simple generators that produce an audio square wave and, consequently, a series of harmonics up into the radio-frequency range, but you will frequently want a tunable rf signal generator. A project in this chapter shows how you can build a simple signal generator for yourself that provides a fixed-frequency audio output and a variable-frequency rf output.

You can buy more complicated signal generators that provide a variable-frequency output. These are called **sweep generators**, because the output frequency is swept from one selected frequency to another at a set rate. These are essential for alignment of FM and television receivers and transmitters.

Some even more elaborate signal generators are used by TV servicemen and technicians to generate color bar and dot patterns needed to align color TV receivers.

Signal Tracers

You may want to know if a signal is really present at some stage in an amplifier. Frequently, you can check using a pair of earphones with a high input impedance. It may help if you use a capacitor in series with one of the leads to provide dc isolation. Sometimes, with very low-level signals, you can use an audio amplifier with a high input impedance to listen for signals. If you want to check for rf signals, you can do so—if the signals are modulated in some way—with a diode in series with one lead of your high-impedance headphones or audio amplifier. At other times, this may not be satisfactory, and you will find it necessary to use a radio receiver tuned to the frequency you are interested in. Disconnect any outside antenna from the receiver and use in its place a probe connected to the receiver's antenna terminals through a piece of coaxial cable.

Bench Power Supply

For an occasional bout of troubleshooting, you can get by with some battery holders hooked up to give you 3, 6, 9, and 12V. If you do a lot of building and experimenting, however, you will find that a regulated, variable-voltage dc power supply will come in handy. This will save you the frustration of discovering, after 3 hours of futile checking, that the reason your project doesn't work is because your batteries are dead.

Transistor Checker

You may be able to isolate a problem to a particular stage of a project, and you may direct your suspicion to a particular transistor. The only way to be certain is to remove that transistor and either insert a substitute or run some kind of a test on the suspect part. Substitution is the best way to test whether any part is working, but you may not have two of every component in your project. You can make a simple test of a common bipolar transistor with just an ohmmeter. The resistance from emitter to collector should be at least a kilohm, and more important, the effect of the transistor base-to-emitter and base-to-collector junctions should be very obvious. That is, for a pnp transistor, you should get a low resistance with the negative lead from the voltmeter connected to the base and the positive lead connected to the emitter or collector, and a high resistance with the positive lead connected to the base and the negative lead connected to either of the other two elements. For an **npn** transistor, the indications would be reversed.

You can tell more about a transistor's condition with a **transistor checker**. This instrument tells you the actual dc current gain of the device. There isn't any simple ohmmeter check that will tell you much about the condition of a field-effect transistor, but the BJT-FET transistor checker described in this chapter can tell you about the condition of FETs as well as bipolar transistors.

Dip Meters and Wavemeters

Experimenters who do a lot of work at radio frequencies—particularly hams who like to try out different antennas—find use for dip meters and wavemeters. The original dip meter used vacuum tubes and was called a **grid-dip** meter. Modern equivalents use bipolar or field-effect transistors, which, of course, have other control elements than grids. No matter what the device inside, the idea of all dip meters is the same.

Each consists of a variable-frequency oscillator (vfo) with an LC tank circuit. The inductor in this tank circuit is exposed and may be coupled to another resonant circuit, such as a tank circuit or an antenna. When the dip meter is tuned to the same frequency as the external resonant circuit, some of the energy from its coil is coupled to the external circuit, resulting in a dip of a meter needle as the dip meter is tuned through resonance. This reveals the frequency to which the external circuit is tuned.

Most dip meters are designed so that they can also be used as wavemeters. In this mode of operation, the oscillator portion of the dip meter is cut out, and a radio signal of unknown frequency is coupled to the coil. In this case, the meter shows an increase when the tuned circuit in the meter is adjusted to the unknown frequency.

Oscilloscope

An oscilloscope is really a kind of voltmeter that lets you see what happens to a voltage over a period of time. Oscilloscopes are essential for aligning FM and television receivers, and they are very helpful to hams for assessing the quality of modulation of their transmitters. For experimenters who do not have an interest in these areas, scopes may prove to cost more than they are worth.

At one time or another, every electronic hobbyist has asked, "Can I make an oscilloscope out of an old television set?" The answer is no; here's why: In an oscilloscope, the deflection of the electron beam is controlled by the electrostatic charge on two pairs of plates in a cathode-ray tube. One pair controls the up-and-down motion of the beam, and the other pair controls the side-to-side motion of the beam. Since no current flows between the plates, the drive requirements are very low. In a TV picture tube, the motion of the beam is controlled by a set of coils. There just isn't any practical way to use these coils for deflection at rates other than the ones they were designed for (15 kHz for horizontal deflection and 60 Hz for vertical deflection). Too bad. It was a nice idea. But don't be embarrassed, it has occurred to most of us.

SYSTEMATIC TROUBLESHOOTING

No matter how much test equipment we have, it doesn't do us much good unless we use it in some logical, systematic fashion. The steps summarized below aren't the only way to go about finding what's wrong with a project, but they represent an approach that will work, and one that can be modified to fit nearly any circumstance.

Was the Job Put Together Right?

Take the time to match the circuit, one stage at a time, to the schematic. Use a pencil to mark off each part of the circuit as you check it.

Are the components the proper value? Check the color code on each resistor and the value printed on each capacitor. Do they match the schematic?

Are the right transistors in the right locations? One transistor looks pretty much like another. Check again to make sure Q2 isn't where Q4 is supposed to be, for example.

Are the transistor lead connections right? Remember, transistor base diagrams read from the **bottom**. Have you interchanged collector and emitter accidentally?

How Does the Project Look?

Inspect your solder joints. Do any seem to be cold solder joints? Do some joints have globs of solder on them? This could mean that there is no solder on the joint, just hardened rosin. Also, big globs of solder could cause a short somewhere—look closely.

Are there any lost nuts, lockwashers, or bits of clipped wire that could cause a short? Pick the equipment up, turn it upside down, and shake it vigorously to get rid of any flotsam and jetsam. If you are afraid that turning the equipment upside down and shaking it will break something, you haven't built your project strong enough.

Are there any items of mounting hardware that could be causing a short? Could they short when the covers are screwed on? If you have any feedthrough insulators or binding posts that must be insulated from chassis or panel, are they shorted out?

Did your soldering iron burn through the insulation on any wires, creating a short-circuit hazard?

How do your components **look**? Are the exterior surfaces of your capacitors smooth, or are they bubbled and darkened from excess heat? Do the color bands on your carbon resistors show signs of discoloration? If a resistor looks bad, tap it with the blade of a screwdriver. A resistor that has failed because of too much current passing through it will often break in two when you tap it. Inspect chokes and power transformers. Excessive current causes heat that may cause potting compound to liquefy and run out. Or it may cause lacquer to overheat, making the part smell burnt. The windings on rf chokes subjected to too much current may be visibly burnt. Usually, when transistors fail, there is no visible sign, but sometimes they may heat up and literally explode. Vacuum tubes that have lost their vacuum may show a whitish deposit inside. If you look at vacuum tubes that are operating and see a bluish discharge inside, the tubes may be gassy or operating under too high a voltage. (Voltage regulator tubes are another story. They're **supposed** to glow. Some glow orange and some glow blue.)

Is DC Power of the Right Polarity Going Where It Should?

Disconnect the dc supply from the equipment. What is its output voltage? Reconnect the dc supply. What is its voltage now? Does the equipment cause the supply voltage to fall below the required level? If it does, is this due to a short in the equipment, or to a supply with inadequate current capacity?

If the fuse blows when you apply power, you are in luck—all you have to do is to find out why it blows. This isn't a facetious statement; a gross short circuit is more easy to locate than a more subtle failure.

With the on-off switch off, measure the voltage across its terminals. The voltage should equal the full supply voltage. If you can't get a reading, there is an open somewhere between the supply and the switch, or the switch and ground. Now turn the switch on and again attempt to measure the voltage. The potential drop across the switch should be zero. If you read the full supply voltage, the switch is broken.

If the project has an ac supply, measure the voltage at the output of the transformer. Is it what it is supposed to be? If it isn't, the supply voltage may be down, the transformer may be defective or mislabeled, or the current drain of the circuit may be too great. The last case indicates, among other things, that the fuse to the transformer is not working or has the wrong value. Measure the output of the filter with the circuit disconnected. You should get a value around 1 to 1.4 times the nominal value of the transformer output voltage (unless you are measuring a voltage multiplier circuit), depending on the size of the bleeder resistor. Leave the circuit unconnected and switch the meter to the ac voltage scale. Any evidence of a substantial ac signal at this point suggests an open filter capacitor. Now reconnect the circuit to the power supply and again measure the dc output voltage. It should be only slightly lower than it was in the no-load case.

Turn the power off. Refer to the schematic and check for continuity (zero resistance) between all ground connections. Make sure that the ground bus is connected to the proper (either negative or positive, depending on the circuit) terminal of the power supply.

Leave the power off and check all of the points connected to the other side of the power supply. Turn the power on and make sure that each point is at the same potential. At each transistor collector, FET drain, or vacuum-tube plate, measure the voltage at each end of the load resistor (if one exists). In general, if there is no change in voltage across the load resistor, the transistor (or other active device) is open-circuited. If the entire supply voltage drops across the load resistor (or across the load and bias resistors), the active device is shorted out.

What Is the Condition of Each Stage of the Circuit?

If your project has an audio output to a speaker or an earphone, use a signal generator to apply an audio tone to the input of the device. Can you hear anything? If you do not have a signal generator, use a battery and a pair of wires to apply a dc voltage to the speaker or earphone. You should hear a click each time you make or break contact if the device is functioning.

If your project has a lamp, light-emitting diode, meter, or other optical device as its output, remove the device from the circuit and check to see whether it is working. Incandescent lamps and LEDs can be tested with an ohmmeter. Incandescents should read a dead short; LEDs should have low resistance in one direction and high in the other, just like other diodes. When the resistance is high, the negative lead of the meter is connected to the cathode of the diode. Gas discharge tubes such as neon bulbs and xenon flash tubes cannot be tested easily except by inserting them in a circuit you know is good. Incidentally, the firing voltage of a neon bulb will increase over a period of time if the bulb is not exposed to light.

Do not check a meter with an ohmmeter. If the questionable device is an ammeter, select a combination of voltage and resistance to give a known current in the meter's range, and use this to verify proper meter operation. If the suspected device is a voltmeter, find a known voltage in the proper range and measure it with the voltmeter. Do not apply ac to a dc instrument.

If your project uses a digital output device with an integral decoder-driver, first check all of the voltage levels at the **lamp test, blanking, ground,** and V_{cc} terminals. The proper levels for these are covered in the Randomizer project. If all of these are as they should be, disconnect the A, B, C, and D inputs to the decoder, and input these manually, observing the effect on the display. Figure 4-1 shows the output of a 7-segment display in response to all binary inputs from zero to 15.

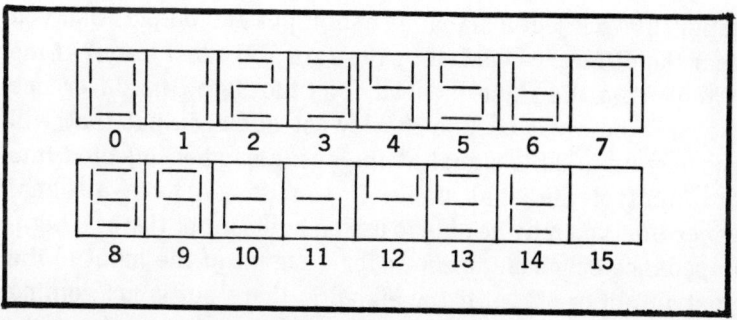

Fig. 4-1. Output of a 7-segment display in response to bcd inputs from zero to 15. Outputs higher than 9 may cause you to mistakenly think that some segments aren't working.

If your project is a signal generator, vfo, exciter, or transmitter with an rf output, is your antenna or dummy load a satisfactory match for the output stage? Has the output been somehow short-circuited? If you are using a radio receiver to look for your signal, is it possible that the receiver is too closely coupled to the output of your project? This could result in blocking the receiver. Also, check again to make sure that the receiver is tuned to the same band as the anticipated signals from the project. If you get many different signals, do not assume that they are all in the output of your project. It is likely that a lot of these are birdies, spurious signals created inside the receiver by its own heterodyning action. You can get rid of most of them by moving the receiver farther away, until it barely picks up the signal from the project. Then look for the lowest, strongest signal you can find. That will be the true output from your project.

The above steps account for troubleshooting most of the different kinds of output devices you will encounter. The inverter project in this book has one of the more unusual output devices you will come across—a power transformer. In this case, you should first check the electrical continuity of the transformer primary and secondary. Then, disconnect the transistor circuitry, connect the 110V output to the house current supply (better use a fuse and wrap all of the connections to prevent a short), and measure the ac output voltage at the 24V side. Remember to use the ac range of your VOM. If the transistor passes this test, it's okay.

If you have an audio output stage, use an audio signal generator to inject a signal at both input and output. Can you hear the difference caused by the gain of the last stage? If the output from the signal generator is too high, the difference may not be obvious. Use the lowest audio output that will produce an audible signal at the speaker when injected into the input of the final stage. If you do not have a signal generator, you may be able to use an audio amplifier or a high-impedance earphone to check for a signal at the input to the final amplifier stage. If the signal is there but is not coming out the other end, you've isolated the bad stage.

If you have an rf output stage, check first to make sure that an input signal of the proper frequency is present. Use a small coil of wire connected to the antenna terminals of a receiver capable of covering the frequency band of interest. You may remove the active device in the output stage entirely or disconnect one of its leads to make sure that it is not affecting the signal. If the input signal is present and the output stage has a tuned tank circuit, check first with a dip meter, if you have one, to verify that the tank circuit can be tuned to the required frequency. Otherwise, connect a milliammeter in series with the device. (In tube circuits, connect it in the cathode lead, for safety.) Observe the current as you tune the tank circuit through resonance; there should be a pronounced dip in current. If there is not, it is possible that the active device has internal capacitances that prevent the circuit from resonating. If this was a potential problem, the kit manufacturer or the author of the article should have some instructions for neutralizing the device. If this is your own design, you'll find an extensive discussion of neutralization in the "Radio Amateur's Handbook."

If you have a digital output stage with a separate decoder-driver IC, follow the procedures described above for digital output displays with integral decoder-drivers.

If your output stage contains an SCR or triac, you'll need a dc supply with an output voltage of 50V or more, an adequate load resistor, and some external means of triggering the gate of the device. (A wire connected to the minus supply is fine, but be careful not to lay it down where it could short out the supply.) Connect the dc voltage across the device in both directions without triggering the device, and observe the voltage drop. The full supply voltage should appear across the device each time. Then trigger the SCR or triac. The voltage drop across a triac should fall to nearly zero, regardless of the polarity of the applied voltage. An SCR, however, should show a high resistance, regardless of triggering, when the positive end of the supply is connect to its cathode and the negative to its anode, and a low resistance, after triggering, when it is forward-biased by the supply.

In some projects, including the inverter in Chapter 2, the output stage is an oscillator. These cases are treated under the heading, "Is the Oscillator Oscillating?"

If the output stage is functioning properly, you will have to work backwards through the preceding stages. For each type of circuit, follow the same procedure that you used for the output stage.

Is the detector, mixer, or modulator working? In the **modulator**, the inputs to the circuit are a radio-frequency carrier and an audio (or composite video) signal containing information to be impressed on the carrier. An ordinary modulator fed with a single tone provides an output consisting of the audio signal, the original rf carrier, and two additional rf signals, or **sidebands**—one with a frequency equal to the carrier frequency **plus** the audio frequency, and one with a frequency equal to the carrier frequency **minus** the audio frequency. The amplitude of the sidebands is proportional to the audio amplitude. Another kind of modulator, called a **balanced** modulator, suppresses the carrier frequency so that only the sidebands and audio signal are present in the output. Usually, there is a simple filter to eliminate the audio in both kinds of modulators. Single-sideband transmitters use another filter to suppress either the upper or lower sideband.

In the **mixer**, used most often in superheterodyne radio receivers, one input signal is the actual radio signal received at the antenna. This consists of the carrier and both sidebands, or in the base of a single-sideband transmission, just one sideband or the other. Another input to the mixer is the signal from the receiver's local oscillator, which is tuned to a frequency above or below the frequency selected by the antenna tuned circuit. The amount by which this frequency differs from the input signal frequency is called the receiver's **intermediate frequency** (i-f). The output from the mixer consists of all the input frequencies, the intermediate frequency, and two sidebands on either side of the intermediate frequency. These sidebands bear the same relationship to the i-f that the original sidebands bore to the rf carrier. Usually, mixers have sharply tuned output stages that filter out the rf and local oscillator signals.

In the **detector**, the input is just some rf or i-f carrier plus one or two sidebands. The detector output contains carrier and sidebands, and the audio signal that originally produced the sidebands. Generally, the detector output is bypassed (filtered) to suppress the rf signals and leave just the audio.

We will use the same technique for troubleshooting modulators, mixers, and detectors. Refer to the discussion of each mode above, and determine what inputs and outputs should be present. Then use a suitable device to check for them. Without a device called a **spectrum analyzer**—which not too many people ever get to see a specimen of—you will not be able to observe sidebands directly, but you will be able to check for rf signals using a separate receiver tuned to the proper frequency. If you can hear the audio in the output of the second receiver, the sidebands are there.

You can also check for the presence of the i-f even though you do not have a receiver that tunes that low, as long as your receiver has the same i-f. Just make a coil of several turns of wire and connect this to your probes. Place the coil near the tube or transistor that serves as the first i-f amplifier in the receiver, and the signal should be audible.

Audio frequencies are most easily checked by means of an audio amplifier or a high-impedance earphone connected to the appropriate terminals of the detector, mixer, or modulator.

If the appropriate signals are present at the appropriate terminals of the circuit, you can dismiss the detector, mixer, or modulator as the source of your troubles.

Check audio oscillators with an audio amplifier or a high-impedance earphone across the output. Use a probe connected to the antenna terminals of a suitable receiver for rf oscillators. If you cannot observe any signal, disconnect the oscillator from the preceding stage and check again. If you get oscillations now, something in the next stage is loading down the oscillator.

It is an unfortunate fact of life that sometimes, in some circuits, one tube or transistor of a given type will not oscillate, when most others of that type will. If you are down to the hair-tearing stage, try substituting another device of the same type. The odds for success aren't great, but you may find that this solves your problem.

Nonsinusoidal oscillators like the multivibrator may be checked in the following way: First, apply power and determine which transistor is conducting and which is cut off. The voltage drop across the conducting transistor will be very low, while the drop across the cut-off transistor will be nearly the full supply voltage. If both transistors are conducting or both are cut off, one or both are bad. If you have determined that one is indeed in the **on** state and one is in the **off** state, try to get the multivibrator to flip into the opposite state. Do this by shorting out the resistor that is connected to the base of the nonconducting transistor (Fig. 2-17). If you are unsuccessful, suspect a failure of either the conducting transistor or the capacitor connecting its collector to the nonconducting transistor's base. If the multivibrator "flips," but "flops" right back, the characteristics of the two transistors are too far apart; one of them will have to be replaced. If the multivibrator "flips" and stays "flipped," the capacitors you are using have too much leakage.

If a nonsinusoidal oscillator using an SCR or triac triggered by a neon lamp or diac refuses to oscillate, first check the SCR or triac as outlined above. If that checks out okay, the problem is either a faulty neon lamp or diac, or too low a voltage on the high side of the lamp or diac. This could result from a bad power supply—which you should have detected before this—or from an open resistor or leaky capacitor in the timing circuit.

Which Components Are Causing the Problem?

Let's assume that you have followed the steps described in the preceding paragraphs and have isolated a faulty stage. Your next task, if you are building with discrete components, is to isolate and replace the faulty component in that stage. If you are using ICs, your problem is solved, since you simply have to replace the faulty IC. Don't be too hasty about replacing that IC, however; give it one last functional check when it is out of the circuit, to see whether some problem with another stage hasn't been giving you a false indication. And try to see why the IC failed. Was its supply voltage too high? Was it wired incorrectly? Is the current drain of the next stage excessive? Could the IC have been damaged by excessive heat during installation?

Let's look at some of the tests we can perform on discrete components.

Resistors. As noted earlier, you can spot some burned-out resistors by discolored code bands. If there is any doubt, though, it is best to measure the resistance with an ohmmeter. To get an accurate reading, it frequently will be necessary to disconnect one end of the resistor from the circuit.

Capacitors. Some capacitor failures you will be able to **hear.** In a radio receiver or audio amplifier with a 110V ac supply, an open-circuited filter capacitor will betray itself by a loud hum audible from the speaker and independent of volume control setting. Short-circuited capacitors are revealed by a simple resistance measurement, but be certain the capacitor is disconnected from the rest of the circuit before you decide it is shorted. Be careful also, when dealing with large capacitances, that you discharge the capacitor before you touch it. The only way to check small capacitors that do not appear to be shorted is by means of an instrument like the Capaci-Bridge described later.

Inductors. The easiest test you can perform on an inductor is to measure its dc resistance. On most coils, this should be very low. If you have established that the coil is not open, you still cannot tell whether the coil has the proper inductance without measuring it, and this requires some fairly sophisticated equipment.

If the coil is made from air-wound coil stock, or if you wound it yourself in a single layer, you can calculate its inductance to verify that it is indeed the proper value. Figure 4-2 is a graph that gives the relationship between the ratio of coil diameter to coil length (d/ℓ) and a form factor (F). To calculate an inductance, first measure the length and diameter of the coil and calculate (d/ℓ). Then refer to Fig. 4-2 to find F. The inductance in microhenrys is given by

$$L = Fn^2 d$$

where **n** is the number of turns and **d** is the diameter of the coil.

Semiconductors. Diodes, of course, can be tested with an ohmmeter. Most diodes are made either with the circuit

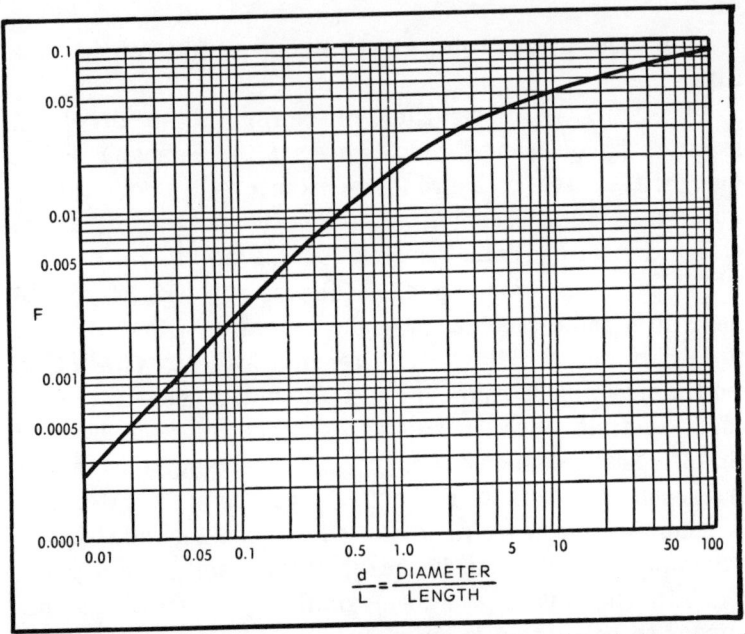

Fig. 4-2. Graph for finding form factor (F), given length and diameter of a single-layer coil. The value of F is then used to calculate inductance.

symbol printed right on them or with a band of paint around the cathode end. In the case of the circuit symbol, the arrowhead is the anode end, and the bar is the cathode end.

Bipolar transistors can be given a simple test using an ohmmeter, as described in the section of this chapter dealing with transistor testers. The BJT-FET transistor checker project provides a simple device that can check both kinds of transistors.

Vacuum Tubes. You can check whether a tube filament is burned out by measuring the resistance between the two base pins connected to the filament. For other tests, there is no substitute for a tube tester. The supermarket variety, known as **emission testers**, are adequate to detect gassy tubes and most common failures. More expensive, laboratory-quality testers, or **transconductance** testers, give more insight into a tube's condition. High-power tubes such as TV horizontal output tubes, transmitting tubes, and audio power-amplifier tubes cannot be tested on any kind of tester for anything but gas or open filaments.

AFTER YOU FIND THE TROUBLE

Once you have isolated the cause of the failure, your troubles are almost over. Before you replace the defective component, ask yourself this: Was it bad when I installed it, or did it fail because of some other defect in a nearby circuit. Don't install that new part until you are certain it will not follow its predecessor down the path to destruction.

Even at this stage, you may still not have met with success. There is no guarantee that there will be only one "bug" in a project. If the equipment still doesn't work, stay calm—go get another cup of coffee—and continue with the troubleshooting procedures I have discussed. Look on the bright side; think of all the potential trouble areas that you have already eliminated.

CAPACI-BRIDGE PROJECT

It is relatively easy to test a capacitor of several microfarads with just an ohmmeter. If you connect the ohmmeter across the capacitor, there should be an initial jump of the needle, after which it should settle back to nearly infinite resistance. There is no such easy test for smaller capacitors. This can make locating a faulty capacitor a difficult job. The Capaci-Bridge (Fig. 4-3) is useful in checking

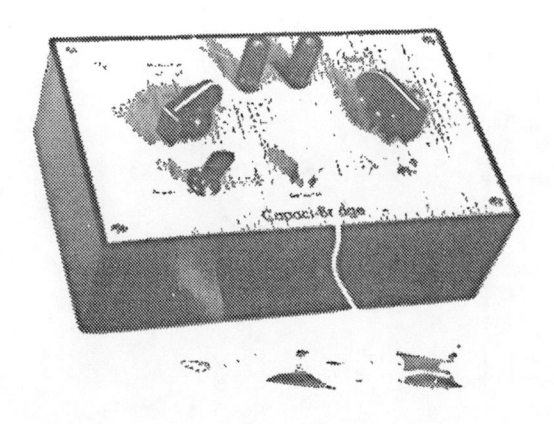

Fig. 4-3. Capaci-Bridge capacitor checker.

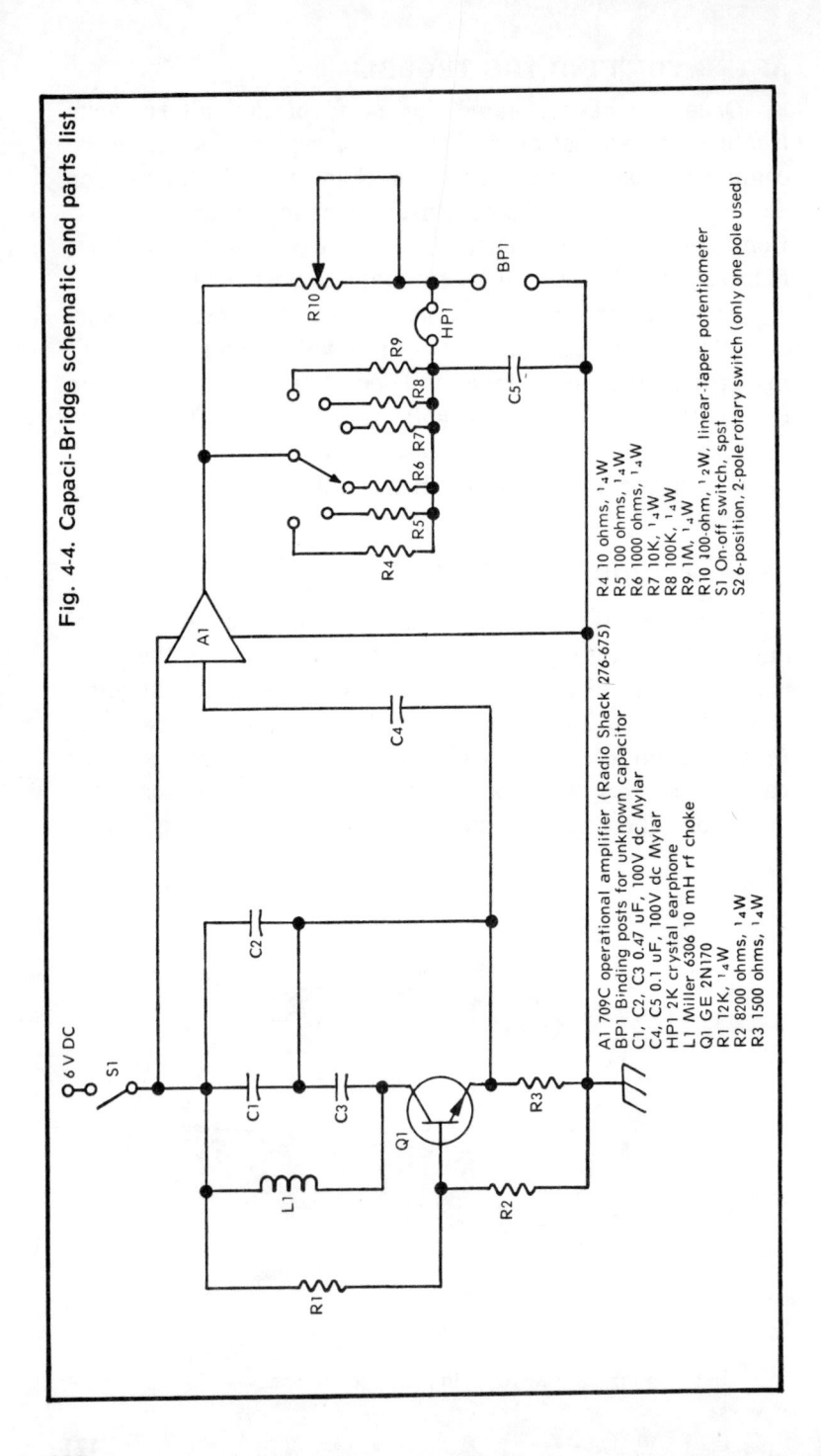

Fig. 4-4. Capaci-Bridge schematic and parts list.

A1 709C operational amplifier (Radio Shack 276-675)
BP1 Binding posts for unknown capacitor
C1, C2, C3 0.47 uF, 100V dc Mylar
C4, C5 0.1 uF, 100V dc Mylar
HP1 2K crystal earphone
L1 Miller 6306 10 mH rf choke
Q1 GE 2N170
R1 12K, ¼W
R2 8200 ohms, ¼W
R3 1500 ohms, ¼W
R4 10 ohms, ¼W
R5 100 ohms, ¼W
R6 1000 ohms, ¼W
R7 10K, ¼W
R8 100K, ¼W
R9 1M, ¼W
R10 100-ohm, ½W, linear-taper potentiometer
S1 On-off switch, spst
S2 6-position, 2-pole rotary switch (only one pole used)

capacitors and in determining the value of capacitors that are unlabeled. The circuit (Fig. 4-4) is basically a one-kilohertz bridge using a high-impedance crystal earphone as a detector.

Capaci-Bridge Theory

The basic bridge circuit is much like the one drawn in Fig. 4-5. Consider the bridge to be just two parallel voltage dividers. One voltage divider is composed of R_a and R_s, and the other is composed of R_b and R_u. The source applies a signal across the pair of voltage dividers, and the detector is placed between the junction of R_a and R_s and the junction of R_b and R_u.

To understand what happens in the bridge, imagine for the moment that R_s is twice as big as R_a. This means that at the junction between R_a and R_s there is a signal with respect to ground that is equal to two-thirds of the signal from the source. If R_u is zero and R_b is any value greater than zero, the signal across the detector will be essentially two-thirds of the source signal. If we gradually increase the value of R_u, the signal

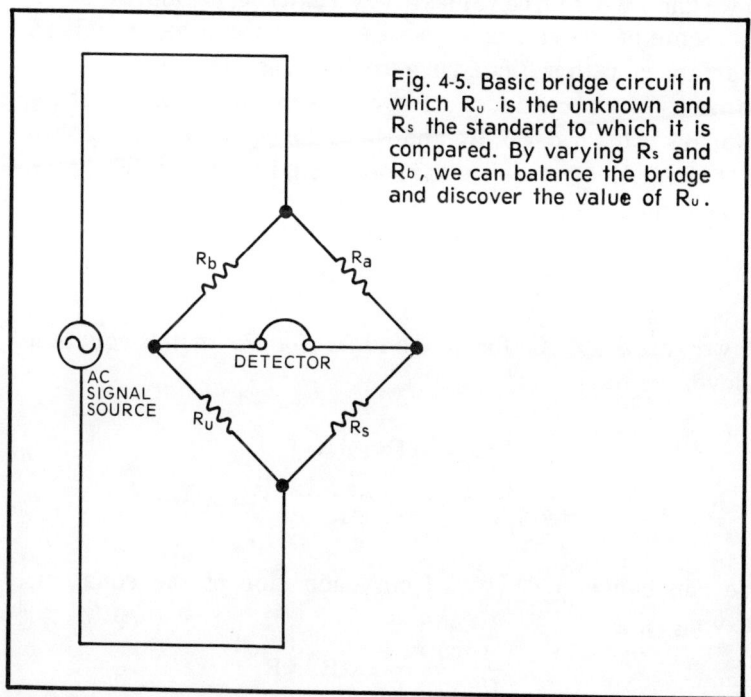

Fig. 4-5. Basic bridge circuit in which R_u is the unknown and R_s the standard to which it is compared. By varying R_s and R_b, we can balance the bridge and discover the value of R_u.

across the detector will gradually decrease. It will reach a minimum value (theoretically, zero) when R_u is twice as big as R_b, because then the voltage at each end of the detector will be the same. If we continue to increase R_u, a signal will again appear across the detector and will increase as R_u increases.

What we are interested in is the situation when the output of the detector is nulled. From the explanation above, we can understand that the value of R_a is to the value of R_s as the value of R_b is to the value of R_u. Stated algebraically,

$$\frac{R_a}{R_s} = \frac{R_b}{R_u}$$

If the value of R_u is unknown, we can solve the equation for R_u in terms of the other resistances.

$$R_u = R_s (R_b / R_a)$$

If we know what three values are, we can find the fourth.

Some of the values in these equations can be capacitive reactances rather than pure resistances. If we deal with capacitive reactances, however, something interesting happens to the order of R_b and R_a in the equation. Remember that capacitive reactance is proportional to the **reciprocal** of capacitance.

$$X = \frac{1}{2 \pi f C}$$

If we substitute X_u for R_u, and X_s for R_s in the equation above, we have

$$X_u = X_s (R_b / R_a)$$

or

$$\frac{1}{2 \pi f C_u} = \frac{1}{(2 \pi f C_s)} (R_b / R_a)$$

We can cancel $1/2 \pi f$ from each side of the equation, leaving

$$\frac{1}{C_u} = \frac{1}{C_s} (R_b / R_a)$$

If we move the capacitances to the numerator, we get

$$C_u = C_s (R_a / R_b)$$

This tells us some interesting things. It tells us that the action of the bridge is not influenced by the frequency of the source and that an unknown capacitance can be compared to a known, standard capacitance times the ratio of two resistances.

You might ask whether the same theory could be applied to inductors. It could, except that in the case of inductors, some practical considerations get in the way. The values of inductance we are most interested in measuring are in the microhenry range. If we use an audio-frequency generator for our bridge, the inductive reactance of these coils is very small, on the same order as the lead resistances in our bridge. Obviously, then, we cannot measure small inductances with an audio-frequency bridge. To successfully measure small inductances, we need a **radio**-frequency bridge. However, this creates much more potential for design problems. The shielding requirements, in particular, require considerable time to solve. That is the reason this project is a capacitance bridge rather than an inductance-capacitance bridge.

Capaci-Bridge Circuit

The bridge circuit is described above. Resistor R_a is a 100-ohm potentiometer, and R_b is a switch-selected fixed resistor of 10, 100, 1,000, 10,000, 100,000, or 1,000,000 ohms. Capacitor C_s is 0.1 uF. This combination of values lets us measure unknown capacitances from a few picofarads to one microfarad.

The oscillator (see Figs. 4-4 and 4-6) is a 2.5 kHz Colpitts type. To keep the bridge from loading down the oscillator, a 709 operational amplifier is used as a buffer. These operational amplifiers are compact, cheap, and have high input and output impedances. The operational amplifier is installed on a printed circuit board (Fig. 4-7). Only one input is used, and the frequency compensator terminals are left open.

The detector is simply a high-impedance crystal earphone. This is simple, cheap, and efficient. The human ear is

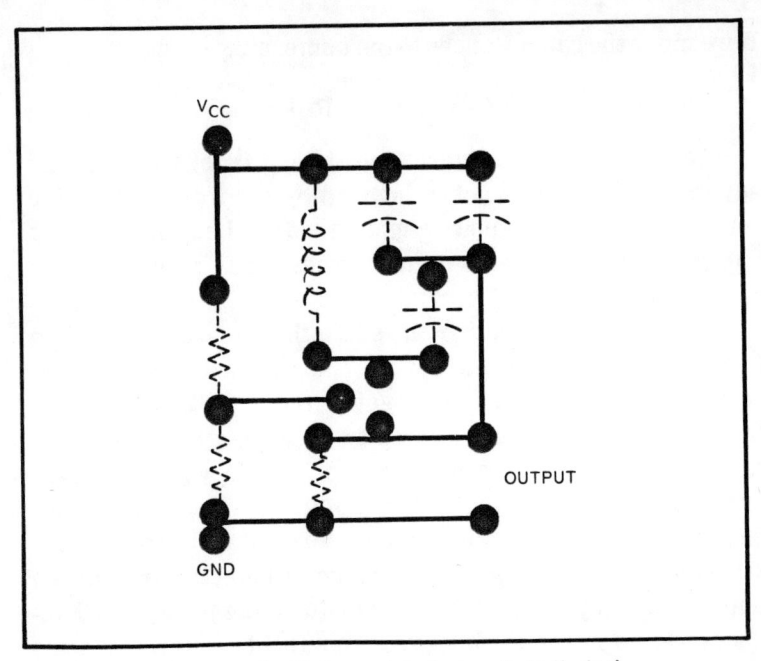

Fig. 4-6. Oscillator printed circuit (full size).

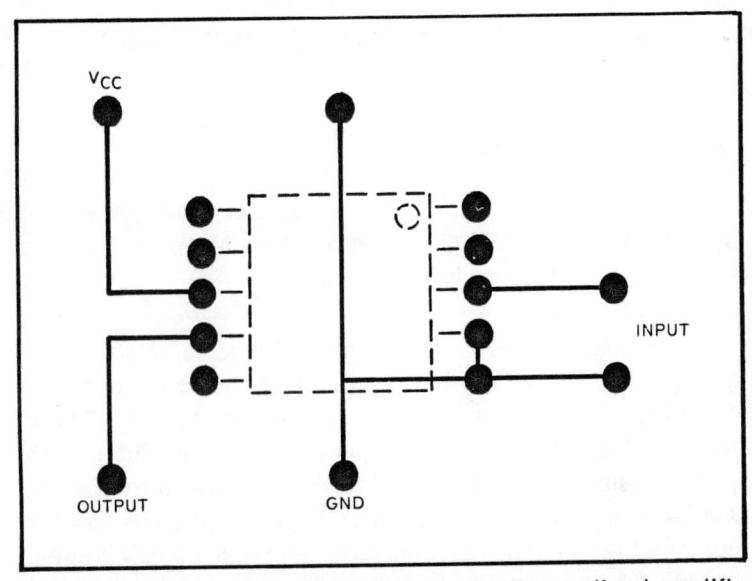

Fig. 4-7. Printed circuit pattern for flat-pack 709 operational amplifier buffer (4 times life size).

very good at detecting nulls. A meter circuit that would be as good would be relatively costly.

Capaci-Bridge Construction

The Capaci-Bridge is housed in a 4¼x7½x2 in. black Bakelite box. Half of a 2-pole, 6-position rotary switch is used to switch the resistors making up R_b. Resistor R_a is a 100-ohm, linear-taper potentiometer. The circuit boards (Figs. 4-6 and 4-7) are mounted on the front panel with right-angle brackets (Fig. 4-8). The photographs show more detail of the construction.

The metal front panel is covered with wood-grain contact paper and labeled with transfer letters. It is no great trick to locate the labels for the multiplier switch, but the VALUE potentiometer presents a more difficult problem.

The problem can be solved by noting that the potentiometer rotates through 315 degrees, between zero and 100 ohms. Knowing this, we can take a piece of paper and make a circle the same diameter as the pointer knob for the pot. We can then use a protractor to strike off division lines every 31.5 degrees. We will need 11 lines in all. Now we can cut the circle out of the paper and position it over the potentiometer shaft

Fig. 4-8. Interior of Capaci-Bridge.

hole in the panel, so that the middle division line (5) is at 12 o'clock. It is an easy task to use the marks on the paper as guides for placing the required numbers.

Capaci-Bridge Operation

Turn on the Capaci-Bridge and insert the earphone in your ear. You should hear an audio signal. (You may not get a signal if the potentiometer is at zero. Pick some point in the middle of its range.) Connect the capacitor to be tested across the UNKNOWN terminals. In each position of the multiplier switch, rotate the VALUE potentiometer through its full range. If the capacitor is good, at some point you will hear the signal in the earphone fall through zero and rise again as you continue rotating the pot. Center the pointer in the middle of the null and read the value. Multiply this times the value shown on the multiplier switch, to find the value of the unknown capacitor.

BJT-FET TRANSISTOR-CHECKER PROJECT

How often have you found yourself in this situation? You've constructed a project according to some plans you've found in a magazine or a book like this, you've applied power, and...and nothing happens.

You begin a thorough logical analysis. Yes, the circuit matches the schematic. Yes, all of the solder joints are solid. Well, how about the components? Are the transistors functioning? How can you tell?

If you have this BJT-FET Transistor-Checker, you can tell in a hurry. It's actually a fairly simple device, nowhere near as complicated as transistors testers can get, but it will test your transistors with about the same reliability as a drug store tester checks your vacuum tubes.

As we said, this is a fairly simple project. All you need is a meter, a light, one transistor, a handful of resistors, and three switches. Yet in spite of its simplicity, it can be one of the handiest items of test equipment you will ever build. If you take the time to duplicate the special meter scale and panel layout shown in Fig. 4-9, you will have an impressive piece of gear to display as well.

THEORY OF TRANSISTOR-CHECKER

The BJT-FET transistor checker performs two measurements on conventional bipolar junction transistors (BJTs) and two go/no-go tests on field-effect transistors (FETs).

BJT Tests

Measuring leakage current from emitter to collector with the base open (Fig. 4-10) is a good check of a transistor's output resistance. This is the first thing to examine. For the most part, a good transistor will cause only a barely perceptible movement of the meter needle. If the leakage current indication moves out of the GOOD region, be very suspicious. Of course, a short-circuited transistor can peg the meter needle.

The leakage test will smoke out a short-circuited transistor, but it cannot tell whether a transistor has opened or not. For that, we need a **gain test**. This not only tells us whether or not the transistor is open, it also gives us an idea of how good an amplifier the transistor is.

The gain we are measuring is the **dc current gain,** or **beta**, of the transistor, generally referred to as h_{FE}. The current

Fig. 4-9. BJT-FET transistor-checker.

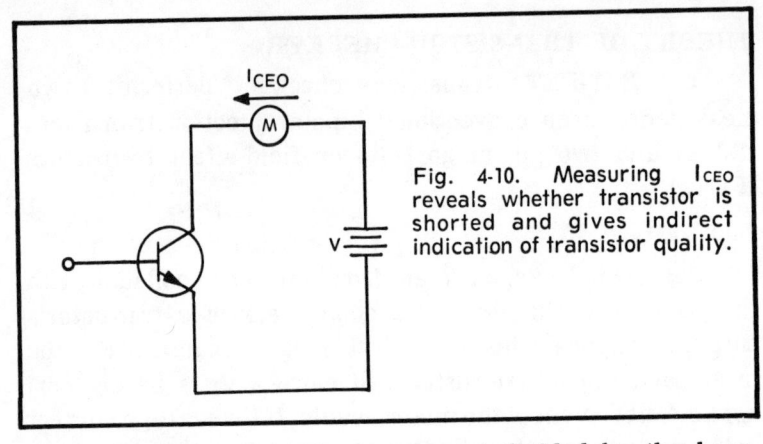

Fig. 4-10. Measuring I$_{CEO}$ reveals whether transistor is shorted and gives indirect indication of transistor quality.

gain is equal to the collector current divided by the base current. In a good transistor, the gain can be anywhere from 20 to over 100, but it averages around 50.

To measure gain, we forward-bias the base of the transistor with a fixed current. In this case, our supply voltage is 9V and we use a 300K resistor in the base lead, so our base current is effectively 30 uA. With gains of from 20 to 100 or so, we can, therefore, expect collector currents of from 0.6 mA (at a gain of 20) to 3 mA (at a gain of 100). Obviously, then, our meter must read from zero to 3 mA.

While it is possible to buy such a meter, it is not as common as the one-milliampere meter used in the BJT-FET transistor checker. To make the meter read full scale when the collector current is 3 mA, the meter is inserted in a current divider (Fig. 4-11) made up of a 5K and 2.5K resistor in parallel. Ammeters have a very low internal resistance (in contrast to voltmeters, which have a very high internal

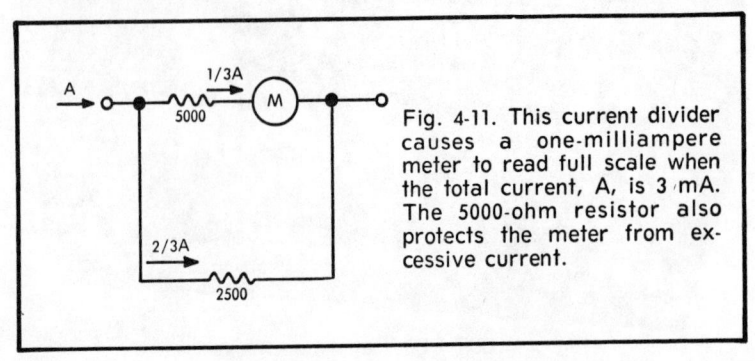

Fig. 4-11. This current divider causes a one-milliampere meter to read full scale when the total current, A, is 3 mA. The 5000-ohm resistor also protects the meter from excessive current.

resistance), so the meter does not affect the current divider circuit. Of the total current flowing through the collector, two-thirds flows through the 2.5K resistor and one-third through the 5K resistor and meter.

This has a very helpful effect from the standpoint of calibration, since the gain can be read directly from the meter scale. How did this happen? Note what occurs when the gain is 100. The collector current in this case is 100 times 30 uA, or 3 mA. This is **full scale** on the meter, or 1. When the gain is 60, the collector current is 1.8 mA. One-third of this, or 0.6 mA flows through the meter, so the needle points to 0.6. You can see that the gain is simply 100 times the meter reading.

In the transistor checker, the meter scale has been modified as described in Chapter 3, and the multiplication by 100 has been incorporated in the scale, so that it reads gain directly.

Field-Effect Transistors

To help understand the tests that the BJT-FET transistor checker makes on field-effect transistors, let's review what goes on inside the FET.

Think of the FET as having a channel from source to drain. There are two ways electricity flows through this channel. If the FET is made with an n-channel, electrons carry the current flow, and the drain is customarily made positive with respect to the source. If the FET is made with a p-channel, so-called **holes** in the crystal lattice carry the current flow, and the drain is made negative with respect to the source.

In either case, the current flow is controlled by means of the gate electrode. For an n-channel FET, the greater the negative voltage (with respect to the source) on the gate, the less current flows in the channel. For a p-channel FET, the greater the positive voltage on the gate, the less current flows in the channel. In either case, the gate voltage can cut off all current flow in the channel, if it is large enough. Figure 4-12 illustrates this effect.

There are three kinds of FETs, each of which can be either n-channel or p-channel. Two of the types of FETs—the junc-

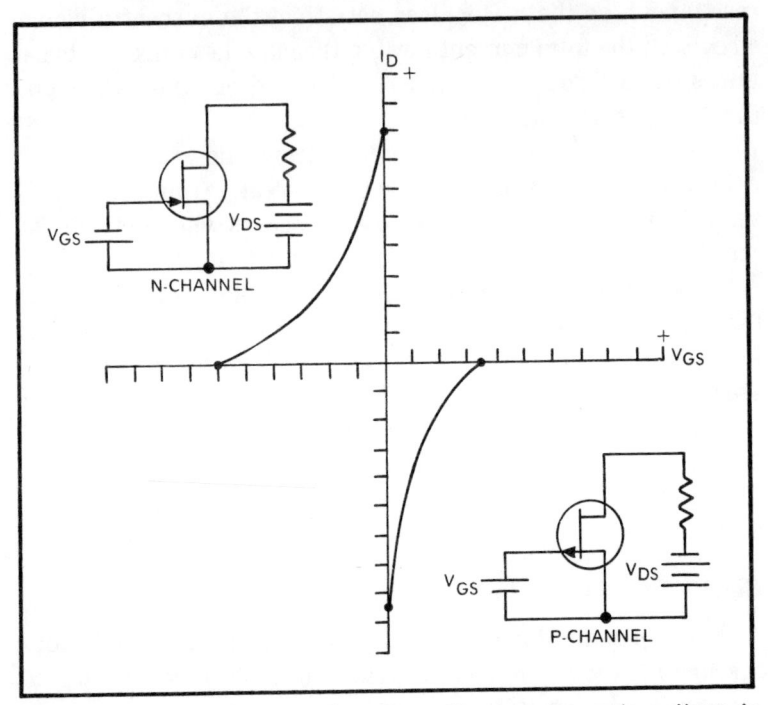

Fig. 4-12. How drain current varies with changing gate voltage in n-channel and p-channel FETs (V_{DS} is held constant).

tion FET, or **JFET**, and the depletion metal-oxide FET, or **depletion MOSFET**—function pretty much alike. In both types, the channel is normally open. For n-channel FETs like these, the gate must be more negative than the source to cut off the channel. Similarly, for p-channel FETs like these, the gate must be more positive than the source to cut off the channel.

The third kind of FET, which may also be n-channel or p-channel, is the enhancement metal-oxide silicon FET, or **enhancement MOSFET.** In this device, the channel is not open unless there is a voltage on the gate. In n-channel enhancement MOSFETs, the gate must be more **positive** than the source for current to flow; current is cut off when the gate is grounded to the source. Conversely, in p-channel enhancement MOSFETs, the gate must be more **negative** than the source for current to flow; current is cut off when the gate is grounded to the source.

If you find all of this confusing, that's not surprising. We're dealing with six different devices and several levels of gate voltage. Table 4-1 should help clarify these matters.

Table 4-1. Biasing of Various FETs.

Channel	Device	Gate-to-Source Voltage		
		Positive	Negative	Same Potential
n	JFET	Conducts	Cut off	Conducts
	Depletion MOSFET	Conducts	Cut off	Conducts
	Enhancement MOSFET	Conducts	Cut off	Cut off
p	JFET	Cut off	Conducts	Conducts
	Depletion MOSFET	Cut off	Conducts	Conducts
	Enhancement MOSFET	Cut off	Conducts	Cut off

FET Tests

The table suggests a way of checking FETs, and this is exactly the test that the BJT-FET transistor checker performs. What we have to do is to apply positive and negative gate voltages to the FET under test and to short its gate to its source. We can make the magnitudes of the voltages large enough to insure that we have complete cutoff or complete saturation in each case. If we can find out what the condition of the channel is for each applied voltage, we will know whether an FET is good or bad.

The current through the FET channel during saturation may not be very great. To get a positive indication, we use a junction transistor (Q1 in Fig. 4-13) as a lamp driver. When the channel is cut off, no current flows through R5 and R6, and the base and emitter of Q1 are at the same potential. When current flows through the FET, R5 and R6 form a voltage divider, and Q1 is biased on, lighting lamp I1. Resistor R7 is selected so that at the lamp's operating current of 60 mA, the voltage across the lamp is 2V.

In the transistor checker, I used an incandescent lamp. However, an even better indicator would be an LED, such as the Radio Shack No. 276-026. This is brighter, and will last many times longer than an incandescent lamp. It is also more

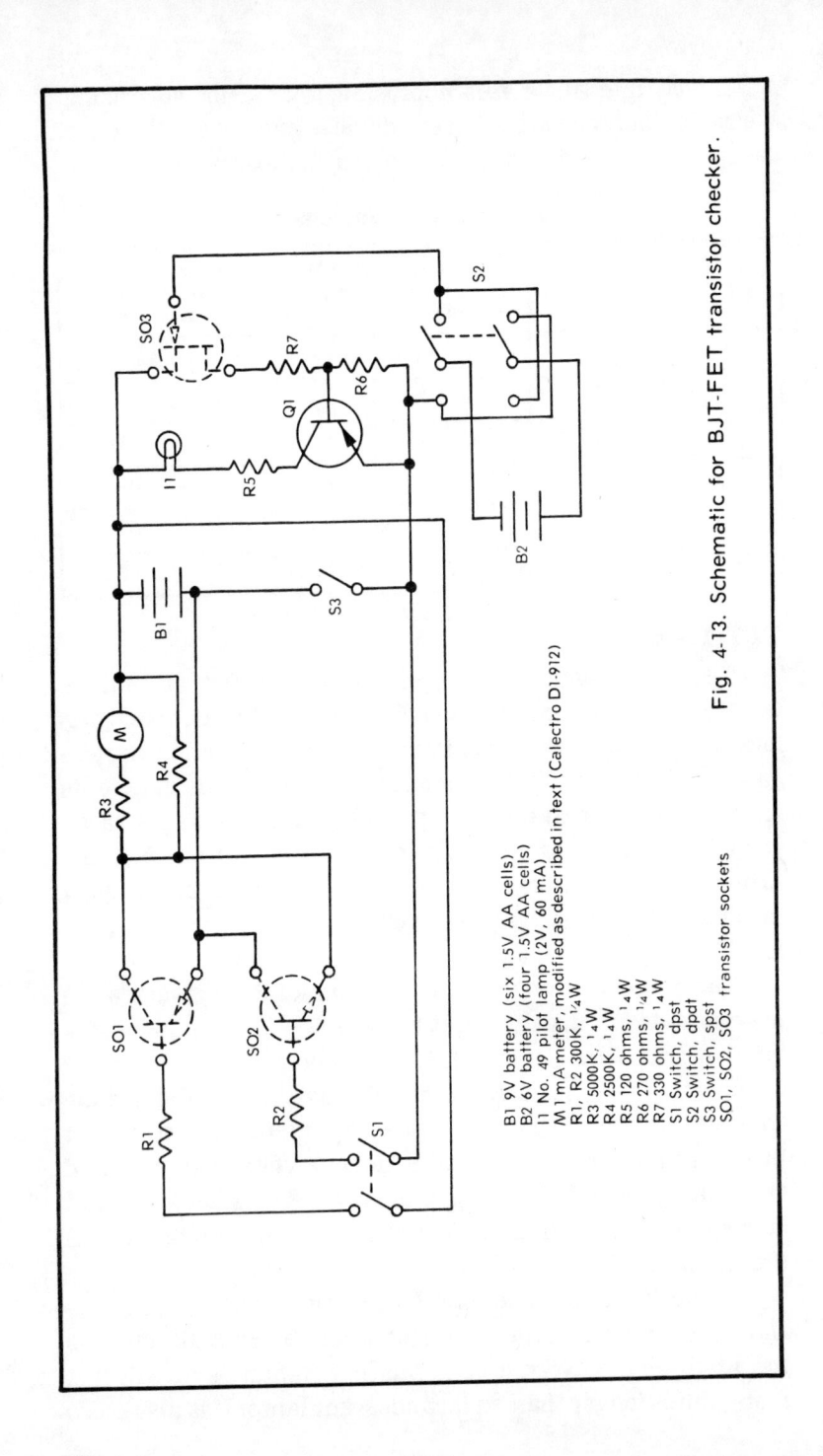

Fig. 4-13. Schematic for BJT-FET transistor checker.

B1 9V battery (six 1.5V AA cells)
B2 6V battery (four 1.5V AA cells)
I1 No. 49 pilot lamp (2V, 60 mA)
M 1 mA meter, modified as described in text (Calectro D1-912)
R1, R2 300K, ¼W
R3 5000K, ¼W
R4 2500K, ¼W
R5 120 ohms, ¼W
R6 270 ohms, ¼W
R7 330 ohms, ¼W
S1 Switch, dpst
S2 Switch, dpdt
S3 Switch, spst
SO1, SO2, SO3 transistor sockets

BJT/FET
Transistor Checker

Fig. 4-14. Copy of meter scale for meter specified in parts list (full scale).

rugged and requires a smaller mounting hole. Remember, the LED is a diode and must have its cathode connected to the ground bus.

Transistor-Checker Construction

The unit in the photos was constructed in a 4-5/16x7x 4-1/16 in. sloping-panel box. The meter scale was modified as described in Chapter 3. Figure 4-14 is a life-size copy of the scale used on the meter. Remember, if you use a different meter, you will have to make your own scale. (This isn't very hard.)

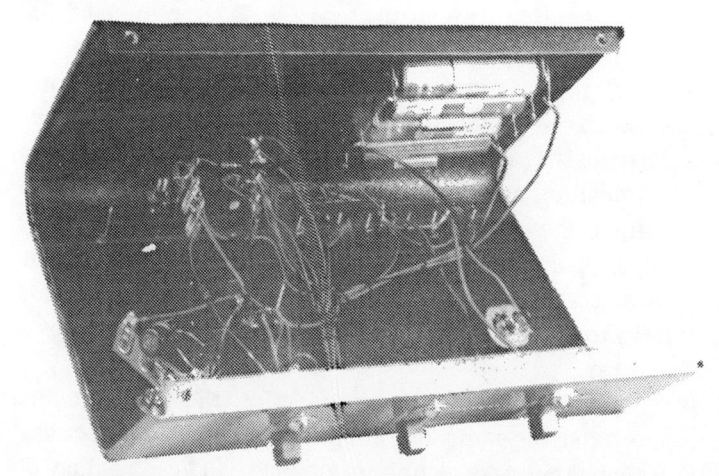

Fig. 4-15. Interior view of BJT-FET Transistor-Checker.

As you can see from Fig. 4-15, two terminal strips are mounted on the meter's terminals. This provides an easy way of mounting the resistors without drilling extra holes in the enclosure for mounting the terminal strips.

There is no such handy location for the terminal strip used for the FET-testing part of the circuit. This is attached to the panel by means of a screw and nut.

In terms of electronic components, the BJT-FET transistor checker could have been assembled within a much smaller enclosure. The large sloping-panel box was used so that a condensed version of Table 4-1 could be included right on the front of the instrument. Most of the abbreviations are obvious. The idea of the channel cutting off or conducting is indicated by means of the binary symbols "0" and "1." A 1 indicates that the channel is conducting; hence, the lamp is lighted. A 0 indicates that the channel is cut off; hence, the lamp is out.

Transistor-Checker Operation

If you understood the section on theory, you should have no trouble with operation of the BJT-FET transistor checker. For bipolar transistors, select the proper sockets—either NPN or PNP—and plug the transistor in. Select the leakage test first, and turn the power switch on. If the transistor is shorted, the meter needle will move up into the BAD region of the leakage scale. If the transistor is not shorted, the needle will move only slightly. In fact, with good transistors you may not even be able to see the needle move at all.

If the transistor is not shorted, move the function switch to the GAIN position and read the h_{FE} of the transistor on the GAIN scale of the meter. Be suspicious of any transistor that reads below 30, and write off any that indicates below 20. For most transistors, you can find the normal range of gain for that type in the transistor handbook issued by the transistor's manufacturer.

To check field-effect transistors, plug the unit into the FET socket. Use the same source, gate, and drain connections for both n-channel and p-channel devices. (The direction of current through the channel isn't important from the standpoint of these tests.) Turn the power on and observe the lamp with the BIAS switch set, in turn, in both positions and in the middle. Then use the table, or LAMP MATRIX, as it is labeled on the unit, to determine whether the transistor is good or bad. In general, if the lamp stays on no matter what position the switch is in, the FET is shorted. If the lamp stays out no matter what position the switch is in, the FET is open.

SIGNAL GENERATOR PROJECT

A good signal generator is a useful troubleshooting adjunct. With such a device (Fig. 4-16), you are able to inject an audio or radio-frequency signal into any stage of a malfunctioning receiver and observe the effect of the signal. The output of a signal generator is a known quantity; it eliminates guesswork in alignment and repair.

The signal generator circuit in this project is the most complicated discrete-component circuit in this book, yet if you carefully duplicate the printed circuit board (Fig. 4-17), you will find that it is difficult to go astray.

One headache with any signal generator project is calibration of the output. How do you correlate dial settings with specific frequencies? In a little while, we'll explain how you can completely calibrate this signal generator with just two items: a conventional broadcast band receiver and an

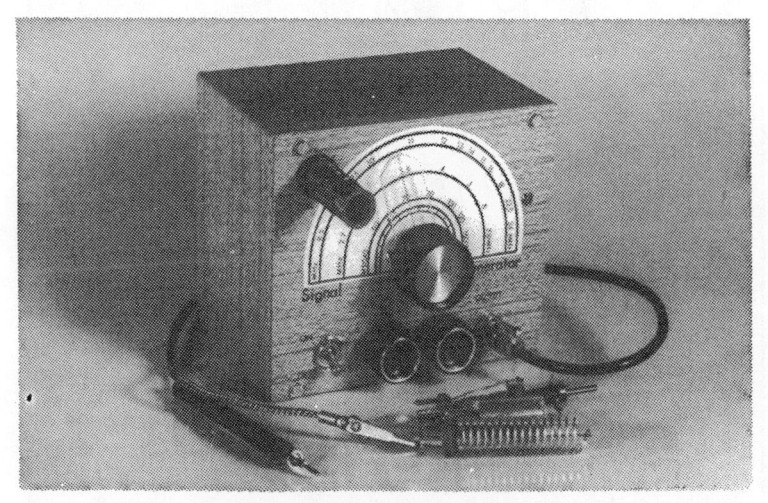

Fig. 4-16. Signal generator.

Fig. 4-17. Full-size signal generator circuit board.

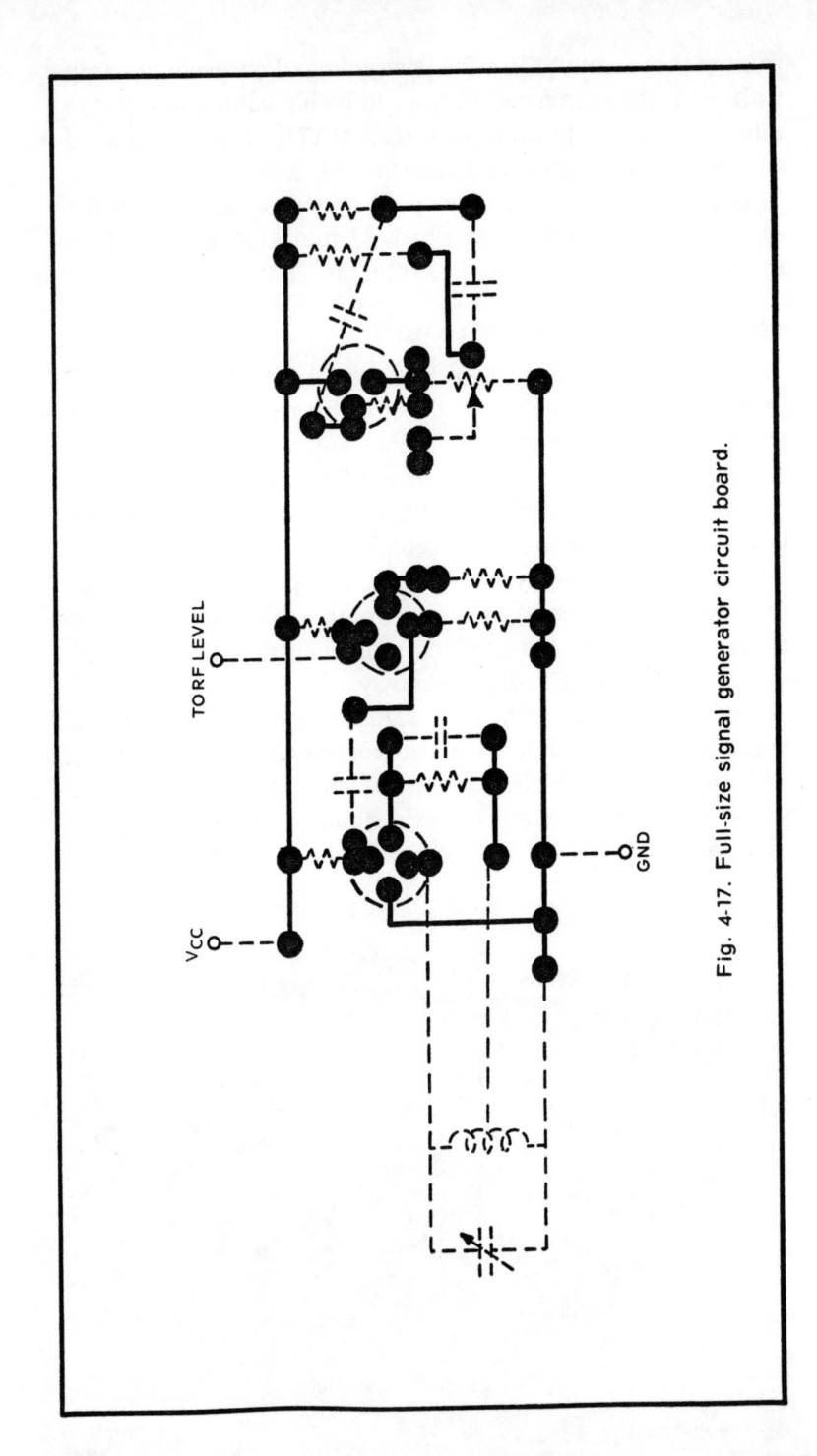

ordinary FM receiver. You can even calibrate two of the bands without an FM receiver if you have a television set handy.

Many commercial signal generators are complicated by elaborate bandswitches. This project avoids that complexity by using plug-in coils. It has most of the other important features of its big brothers, though, including separate audio and rf level controls. The big, easy-to-read dial with the see-through plastic pointer looks professional, but you'll find you can make it yourself in a few minutes.

Theory of Signal Generator

Figure 4-18 shows the circuit of the signal generator. The oscillator is a Hartley type, with a centertapped inductor. This lets us use a standard-broadcast 365 pF variable capacitor for tuning. We used an FET for the oscillator, because it simplifies the design and has good high-frequency characteristics.

Since you will be connecting the output of the signal generator to quite a few different circuits and don't want these circuits to affect the oscillator frequency, the project uses an untuned FET amplifier stage as a buffer. You may notice that this is the same amplifier used in the FET shortwave receiver. This is a good, simple, reliable, and most versatile circuit. The output of this stage is applied to a 1M potentiometer, and the slider on the pot is connected to the center conductor of the output jack.

That accounts for the rf part of the output. We still have to provide modulation for that rf and an audio signal for audio circuits. Bipolar transistor Q3 together with its phase-shift network provides an audio signal of around 400 Hz. We use this signal to modulate the rf signal by applying the output of Q3 to the source of Q2, the output stage. The modulation level control is actually the load resistor for Q3, and the level is controlled by varying the position of the slider on the pot. We can turn Q3 completely off by means of the switch on the back of the audio level control.

Most modulation circuits use a tuned output in the modulator stage or in one of the stages that follow it. This effectively suppresses the audio signal in the output. However, since the output stage of this signal generator is untuned, the

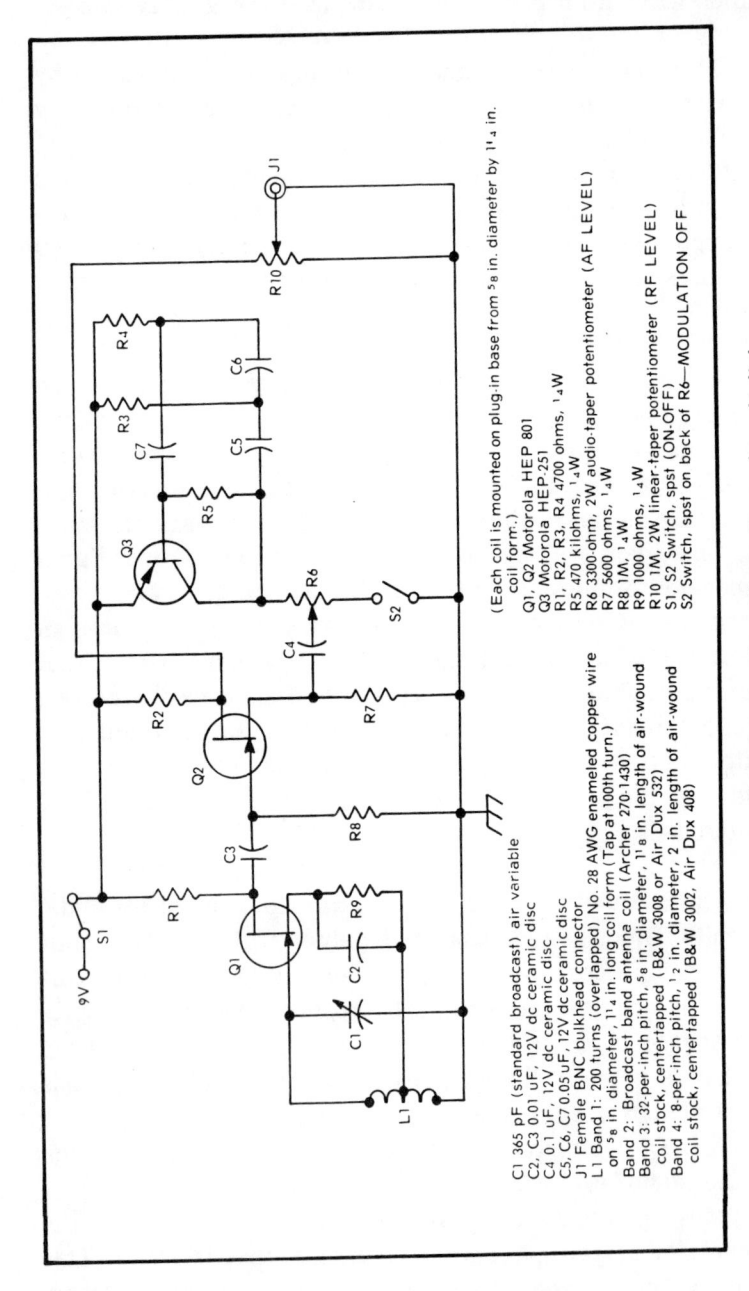

C1 365 pF (standard broadcast) air variable
C2, C3 0.01 uF, 12V dc ceramic disc
C4 0.1 uF, 12V dc ceramic disc
C5, C6, C7 0.05 uF, 12V dc ceramic disc
J1 Female BNC bulkhead connector
L1 Band 1: 200 turns (overlapped) No. 28 AWG enameled copper wire on 5/8 in. diameter, 1 1/4 in. long coil form (Tap at 100th turn.)
Band 2: Broadcast band antenna coil (Archer 270-1430)
Band 3: 32-per-inch pitch, 5/8 in. diameter, 1 1/8 in. length of air-wound coil stock, centertapped (B&W 3008 or Air Dux 532)
Band 4: 8-per-inch pitch, 1/2 in. diameter, 2 in. length of air-wound coil stock, centertapped (B&W 3002, Air Dux 408)

(Each coil is mounted on plug-in base from 5/8 in. diameter by 1 1/4 in. coil form.)
Q1, Q2 Motorola HEP 801
Q3 Motorola HEP-251
R1, R2, R3, R4 4700 ohms, 1/4W
R5 470 kilohms, 1/4W
R6 3300 ohm, 2W audio-taper potentiometer (AF LEVEL)
R7 5600 ohms, 1/4W
R8 1M, 1/4W
R9 1000 ohms, 1/4W
R10 1M, 2W linear-taper potentiometer (RF LEVEL)
S1, S2 Switch, spst (ON-OFF)
S2 Switch, spst on back of R6—MODULATION OFF

Fig. 4-18. Signal generator schematic and parts list.

output contains not only the modulated rf signal but the audio signal as well, so that this is both a variable-frequency rf and a fixed-frequency audio signal generator.

Signal Generator Construction

All of the small components are mounted on the printed circuit board (Fig. 4-17). The board—along with the tuning capacitor, coil socket, switch, pots, and output jack—are mounted on the front panel. The output jack is a BNC female bulkhead connector. A less expensive type of connector could have been used; for example, an RCA-type phono jack. However, the BNC is a connector that stays mated until you remove it on purpose, and it stands up better than a cheaper connector in use.

There are four plug-in coils. Each is mounted on the 4-prong base from a ⅝ in. diameter plug-in coil form. The two high-frequency coils are made of air-wound coil stock, as noted in the parts list (Fig. 4-17). One of these pieces of coil stock has a very fine pitch, which necessitates a special technique for making the centertap connection. Figure 4-19

Fig. 4-19. In a variation of the trick used for the FET-set coil, one turn of the band 3 oscillator coil is bent inward to permit soldering the centertap connection.

Fig. 4-20. The band 2 coil is a commercial loopstick antenna. Leads go to three of the pins on the 4-pin plug.

shows how one loop of coil in the center has been bent inward so that it is separate from the adjacent turns. This allows the center conductor to be easily soldered.

For the broadcast frequencies, we took advantage of the ready availability of prewound loopstick antennas made for replacement use in broadcast receivers. Since these are designed for use with 365 pF variable capacitors, they are ideal for our signal generator. As Fig. 4-20 shows, the loopstick is supported by solid wires which make connection to the base plug. A few drops of cement provide additional support.

The fourth coil is necessary to obtain the low 455 kHz broadcast band i-f signal. This is the only coil that uses the form with the plug-in bases. The coil is composed of 200 turns of No. 28 enameled copper wire, wound in layers over the 1¼ in. length of the coil form. To wind the coil, unscrew the base and drill a small hole in the side of the form. Pass one end of your wire through this hole and start winding. The best way to wind is to keep a constant tension on the wire and wind it up as if you were reeling in a fish and the coil form were your reel. When you reach the 100th turn, twist a big loop in your wire to serve as the lead from the centertap, and continue winding. When you finish, use cement or nail polish to hold the windings in place. When the cement is dry, you can screw the base plug back onto the coil form and make all of your connections to the pins on the base. Be sure to scrape the enamel coating off the ends of the wires before you attempt to solder them. Check with an ohmmeter to make sure you have good connections between each of the pins and the ends and center of the coil.

The outside of the utility box and the front panel are covered with contact paper, as in several other of the projects in this book. There is an important difference, however, in the way the contact paper is applied here. A large part of the front panel is given to the dial, and this must be prepared before the contact paper is applied. If we waited until after the contact paper was on and simply glued the dial scale in place, the edges of the dial scale would be very obvious right from the beginning, and after a while, they would being to curl up. To avoid this, the dial paper is applied first, and the contact paper is placed over it. This guarantees that the edges of the dial will remain neatly in place.

To make the dial, first glue a sheet of white paper over the entire front of the panel. Then locate the center of the hole for the tuning-capacitor shaft and make a mark at this point. Use a felt tip pen—or better yet, a drafting pen—and a large circle template to draw a series of semicircular arcs at 2, 3, 4, and 4½ in. diameters. The arcs should terminate on a line that passes through the center of the tuning-capacitor shaft hole and runs parallel to the top and bottom edges of the front panel. At each point where an arc ends, extend the arc downward in a straight line about half an inch. You'll need this extra half-inch to identify the scales as kilohertz or megahertz. Finally, draw a line across the bottom of the dial scale.

Now you are ready to cover the entire front panel with contact paper. Lay it down neatly, but don't rub it in place just yet. If the lines on the dial face are good and dark, you should be able to see them through the contact paper. Use an X-acto knife and your circle template to score a line around the outside of the outermost semicircle on your dial scale. Finish up along the bottom with X-acto knife and ruler, and carefully peel away the contact paper covering the dial. The contact paper should peel right off, without taking any of the dial scale paper with it.

The dial pointer knob is easily made from an ordinary round knob and a piece of clear plastic from a dime-store plastic box. You can saw the plastic to shape with a keyhole or coping saw, but you will find it just as easy if you score the plastic several times along the line to be cut with an X-acto or other sharp knife and then just break the plastic with your fingers on one side of the score line and a pair of pliers on the other side. You'll find that this way the edges are actually straighter and neater than if you had sawed them. Of course, you'll still need to give them a final dressing on the edge of a file. Use the file to round off all of the corners also.

The index line down the center of the pointer is made by scoring the line with a sharp knife and then applying India ink to the line. Use a tissue to wipe away the ink that beads up and you'll be left with a thin black line down the middle of the pointer.

Glue the pointer to the back of an ordinary round knob and, when the glue is dry, drill out a hole for the capacitor shaft. If the shaft of the capacitor is flatted, be sure that you locate the index line of the pointer diametrically opposite the setscrew on the knob.

The cable for the output probe is RG-58A/U coaxial cable. Figure 4-21 shows how the center conductor of the cable can be extracted through the side of the braid to provide a probe lead and a ground.

Calibration

We'll start our calibration with the broadcast band, since that offers a direct correspondence between frequencies and points on the dial scale of the signal generator. Begin by turning the generator on, with the broadcast band coil plugged in and the tuning capacitor turned fully counterclockwise. Put the probe of the signal generator near the antenna of your broadcast receiver, turn the receiver on, and tune around until you find the lowest signal on the broadcast band. Leave the generator capacitor fully counterclockwise, and adjust the slug in the loopstick until the frequency of the generator corresponds to the lowest dial marking on the broadcast receiver. Now go ahead and match each dial marking on the broadcast receiver with a setting of the generator tuning capacitor. Use transfer numerals to identify each point.

To calibrate the next higher band, you should use the broadcast band calibration marks you have just made and the intermediate frequency of an FM receiver. All FM receivers in the U.S. use an intermediate frequency of 10.7 MHz. If you adjust the signal generator so that it is putting out a signal of 5.35 MHz, the second harmonic of that signal will fall right on the FM receiver's i-f. From our design calculations, 5.35 MHz is **somewhere** within the frequency range of the second coil. To find that frequency, bring the generator probe near the FM receiver's antenna, turn the receiver on, and tune for signals. The signal generator is rich in harmonics, so there will be a lot of signals, but there will be only one that will be independent of the FM receiver's dial setting. When you find that signal, the generator is tuned to 5.35 MHz.

Well, that's just one frequency, you say. How do you calibrate the entire dial? There's a trick. If you grind through the algebra, you will discover that no matter what the inductance, the frequencies corresponding to two settings of the tuning capacitor will always fall into the same ratio.

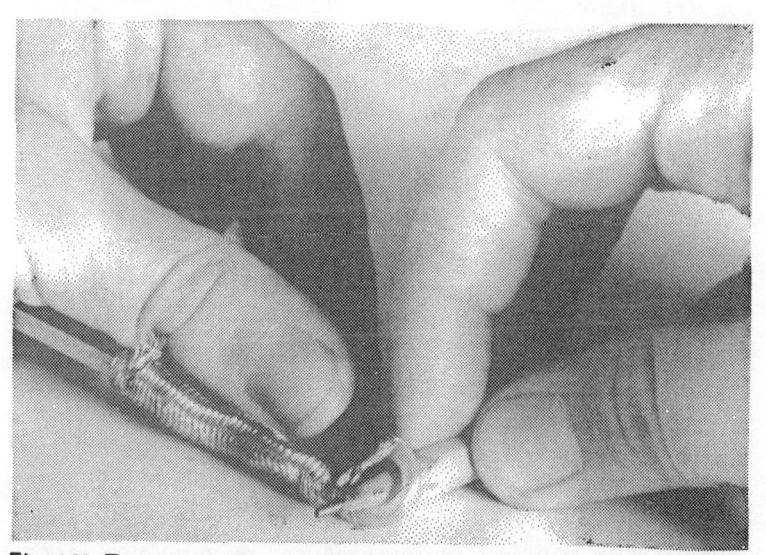

Fig. 4-21. To separate the center conductor from the coax braid, push the braid back and use a blunt instrument to enlarge the hole near the base of the basket weave. Then extract the center conductor through the hole.

Let's see what that means. Take one setting of the tuning capacitor corresponding to 1400 kHz with the broadcast coil plugged in. Take another setting corresponding to 700 kHz, or one-half the original setting. Now it doesn't matter what the actual values of capacitance are—whatever coil you plug in, the frequency at the 700 kHz setting of the capacitor will always be exactly one half the frequency at the 1400 kHz setting.

Here's how to use this mathematical relationship to calibrate the generator. Once you have found the dial setting corresponding to 5.35 MHz, look down and see what broadcast frequency this setting corresponds to. As an example, let's say it corresponds to 1070 kHz. To find out the broadcast band setting for any other frequency, all you have to do is multiply that frequency by 1070 / 5.35. For example, the broadcast frequency setting that corresponds to 5 MHz is 5 x (1070 / 5.35), or 1000 kHz. Similarly, the broadcast frequency setting that corresponds to 4 MHz is 4 x (1070 / 5.35), or 800 kHz. Naturally, you cannot rely on 5.35 corresponding to exactly 1070 on your generator, so you'll have to check it yourself and make your own calculations.

The high-frequency band is even easier to calibrate, because 10.7 MHz itself appears in this band. Once you find it, we proceed as you did with 5.35 MHz, except this time the multiplier is 10.7 / (broadcast band setting).

You can calibrate this band even if there is no FM receiver available, with a television set. The idea is the same as it was with the FM receiver. All TV receivers made in the last 15 years use an i-f of 45 MHz. (Earlier ones used 21 MHz, but there was too much interference from the 15-meter ham band.) Half of 45 MHz is 22.5 MHz, a frequency that will fall within the range of the high-frequency coil of the signal generator. As before, connect the generator probe to the receiver antenna and search for a signal that is independent of channel selector setting. It's interesting to watch the effect of the generator. As before, connect the generator probe of the generator's output very easily, because the audio modulation produces a bar pattern like the one shown in Fig. 4-22. This is often more obvious than the audio signal, since the video detector responds to AM, and the TV audio detector responds to FM.

If you experiment, you will note that harmonics from the midband (2.7-7.8 MHz) coil of the generator will come through the TV i-f. Unfortunately, however, these are such high-order harmonics that the fundamental frequencies are very close together and cannot be differentiated well enough to allow us to calibrate the midband coil with the television receiver.

There is one more calibration to make, the AM broadcast i-f of 455 kHz. It would have been nice if this could have been included in the calibration of the broadcast band frequencies, but it lies just too far below the broadcast band to permit us to do so. This is the reason for the fourth coil, the one we had to wind by hand. The procedure for finding 455 kHz is the same for the other intermediate frequencies: Look for a signal that is unaffected by the dial setting of the broadcast receiver. You can calibrate all of the other frequencies on this band, but this was not done in the case of the generator in the photos.

Fig. 4-22. The effect of the signal generator with full audio output is visible on this TV picture. The TV's video oscillator syncs to a harmonic of the audio frequency to produce this bar pattern. The TV is tuned to a blank channel here.

Signal Generator Operation

The functions of the front-panel controls are self-explanatory. Here's how to use the generator to align an AM broadcast receiver.

Turn the receiver on and set the dial to some position where there is no signal. Adjust the output of the signal generator to 455 kHz and turn the audio and rf level controls up full. Clip the ground lead to the receiver chassis or ground bus of the printed circuit and position the probe near the receiver's last i-f stage until you can hear the audio signal in the receiver speaker. Reduce the generator rf level until the signal is barely audible. Adjust the slugs in the second i-f transformer can to produce the maximum signal at the speaker. (It will be easier if you can measure the voltage at the speaker with the ac range of an electronic voltmeter.) Now move to the first i-f amplifier and repeat, peaking the slugs in its transformer. Always operate with the minimum signal generator level you can hear.

After adjusting the receiver's i-f, return the generator to 540 kHz and set the receiver dial to the same frequency. Now adjust the antenna and local oscillator for maximum signal. On older sets, the antenna-tuning adjustment is a trimmer capacitor on the side of the tuning capacitor, adjacent to the large section, and the oscillator adjustment is another trimmer on the tuning capacitor, adjacent to the small section. On more recent transistor sets, the antenna coil and oscillator coil are adjusted by slugs located within the coils. In either case, the antenna adjustment is usually quite broad, and the oscillator adjustment is quite sharp.

After peaking the receiver at the low end, reset the generator and receiver to 1600 kHz and repeat the procedure. If you find that you have to change the antenna and oscillator settings very much, go back to 540 kHz and recheck your settings there. You may have to "split the difference" to achieve optimum performance and dial accuracy across the entire band.

More sophisticated receivers require more sophisticated alignment procedures. Refer to repair manuals written specifically for these receivers for alignment instructions.

Making Successful Substitutions

The best way to be sure that your finished project is going to duplicate the performance of a project described in a magazine or book is to use parts identical to those in the parts list. As you become more comfortable with electronic construction, however, you may become interested in substituting components different from those in the parts list. You may find that it just isn't possible to obtain some parts, or you may want to use some parts that do not cost as much as those in the parts list, or you may have developed a junkbox which you would prefer to raid before going out to buy new parts. In this final chapter, we will investigate the things you must take into account to insure that the substitute parts will not affect the performance of your project.

RESISTORS

Carbon resistors with a 10-percent tolerance are available in the resistances listed in Table 5-1 and in resistances equal to these values times multiples of 10. For example, consider the fifth value listed, 22. You will find that your radio parts store carries resistors of 22 ohms, 220 ohms, 2.2K, etc.—all the way to 22M. You will find 5 percent resistors with values in between these ranges. The tolerances mean that the actual value of the resistor is somewhere between plus and minus so many percent of the resistor's nominal (color coded) value. That is, a so-called 220-ohm, 10-percent resistor will have a value between 198 ohms (220 − 22) and 242 ohms (220 + 22).

You should understand that the resistor manufacturer doesn't have two machines—one making 5 percent, 220-ohm

Table 5-1. Ten-Percent Resistance and Capacitance Values.

1o	18	33	56
12	22	39	68
15	27	47	82

resistors and the other making 10 percent, 220-ohm resistors. Instead, he has several machines, making resistors of unknown values. All of these resistors are funneled to a testing machine where they are measured and sorted. It's like an electronic sieve separating different sizes of electronic gravel.

The resistors are labeled according to their measured values. All resistors between 176 and 264 ohms are painted with the red-red-brown color bands of the 220-ohm resistor. Those that have a value between 209 and 231 ohms get the additional gold band that signifies a 5-percent tolerance. Some of these may actually be 220-ohm resistors. Of the others, those that measure between 198 and 209 ohms and between 231 and 242 ohms are painted with the silver band that designates a 10-percent tolerance. Note that **none** of these 10 percenters will **ever** be exactly 220 ohms. Finally, all the resistors left over, those between 176 and 198 ohms and between 242 and 264 ohms, will be left with no fourth band and will be sold as 20 percent, 220-ohm resistors.

The point of this discussion is to demonstrate the rather wide variation that can exist between two resistors with the same nominal values. This suggests that in most cases we can expect to get away with substituting a resistor of the next higher or lower value for a particular resistor called for in a parts list. The best clue is whether the author has indicated the tolerance on the parts list or not. If he has indicated 10 percent or no tolerance, we can be very confident in substituting the next higher or lower value, if that is convenient. If the author has taken the trouble to specify 5 percent tolerance, he probably has a good reason for it, and it would be well to use the identical part that he has specified.

There is a new development that you will be seeing on all resistors made in the United States shortly, if you have not seen it already. American resistor manufacturers are now making all of their resistors in conformance to a new military specification, called MIL-R-39008. These are called **established** reliability resistors and they are identified by a fifth color band, following the silver or gold tolerance band. This fifth band indicates the predicted reliability of the resistor in terms of failures per thousand hours of operation. From your standpoint, as someone who buys one resistor at a time, the fifth band is statistically meaningless—but don't let it throw you the first time you find a resistor with an extra band painted on it.

There are some applications in which you can substitute resistance values quite widely. Figure 5-1 shows the output of a power supply containing a **bleeder** resistor. The function of the bleeder is to discharge the power supply filter capacitors when the supply is turned off, to keep them from presenting a shock hazard. Frequently, a value of 1000 ohms is selected for this resistor. It is quite satisfactory, though, to use a resistor of up to five times this value, if that is easier or cheaper.

Another area in which resistor value is not critical is in the inputs of vacuum-tube or FET amplifiers, such as the ones shown in Fig. 5-2A. In contrast to bipolar transistors, in which current flows in the base circuit, very little, if any, current flows in the control element (grid or gate) circuit of a vacuum tube or field-effect transistor. Consequently, the only function

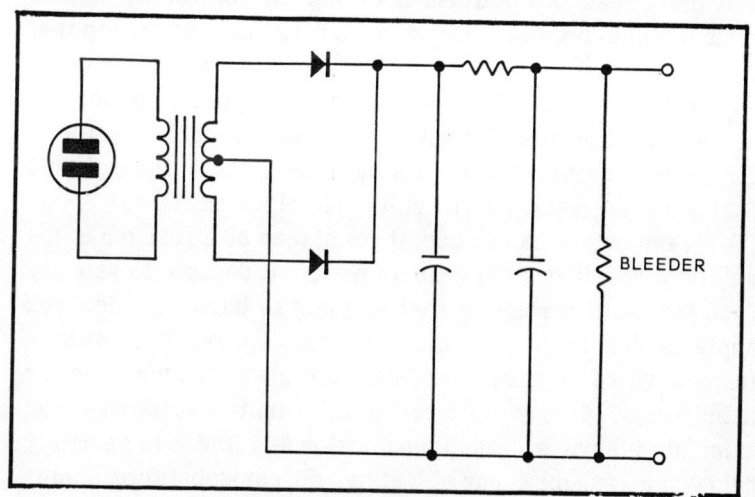

Fig. 5-1. Power supply bleeder resistors may be 1 to 5K. Be sure that the power rating (E^2/R) is adequate.

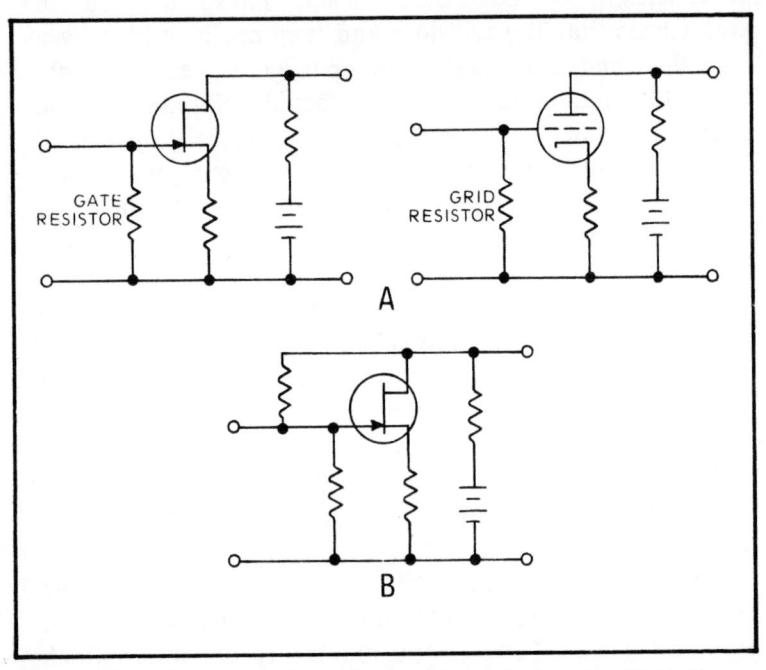

Fig. 5-2. In A, the value of the gate or grid resistor is not critical; it may vary from 10K to several megohms. In B, the resistor is part of a voltage divider that biases the gate, and its value is more critical.

that these resistors perform is to link the control element to ground and to provide a fairly large resistance at the input of the amplifier. In most amplifiers, this resistance can be anywhere from 10K to 10M without affecting performance. Do not confuse this resistor with the voltage divider resistor in Fig. 5-2B, however. In this configuration, the resistor forms part of a bias circuit, and its value is fairly critical.

As important as the actual resistance of a resistor is the ability of the device to dissipate power. It doesn't do you any good to install a resistor that is going to burn up when you apply power to the circuit. You can always substitute a resistor with a higher power rating for one with a lower power rating—e.g., a ½W unit for a ¼W unit—but unless you calculate the power loss in the resistor and find it to be safely below the resistor's power rating, do not substitute a low-power unit for a high-power one. Another thing to remember is that units of different ratings differ in size. If you are copying a printed circuit pattern published in a magazine, be sure that the resistors you use will fit into the space provided for them on the printed circuit board.

Resistors with power dissipation ratings up to 2W are available with carbon composition resistive elements. Larger resistors are generally made by wrapping a ceramic core with resistance wire. Most of these resistors are used in power supply applications, and the method of their construction does not often concern us. However, in applications in which audio or radio frequencies are involved, the inductance created by these windings of resistance wire can create problems. This is not a difficulty you will encounter very frequently, but when you do, you should know that there are two ways you can get around it. One way is to use a large number of carbon composition resistors in parallel to achieve the desired resistance at the required power level, and the other way is to wind your own resistor using what is called a **bifilar** winding. This technique involves taking the whole length of resistance wire needed to make your resistance and doubling it back on itself to make a sort of long, narrow hairpin. When this hairpin is wound on the ceramic core of the resistor, the inductance effect is canceled out, because the currents in adjacent turns are flowing in opposite directions.

The considerations outlined in the preceding paragraphs apply to potentiometers and resistors with sliding taps, as well as to fixed resistors. One additional consideration to take into account when substituting potentiometers is the device's **taper**, the shape of the curve of resistance vs shaft position (or slider position, on slider pots). Figure 5-3 compares linear and audio tapers. The audio taper is tailored to the response of the human ear, which is essentially logarithmic. If you substitute a pot with a linear taper in a volume control application, you'll find it difficult to set a satisfactory level at the low end of its range; it will seem as if a small change in setting will produce an extremely large change in signal level.

CAPACITORS

With respect to choosing the next highest or next lowest value of capacitance instead of the specified value, the remarks under the heading "Resistors" apply equally well to

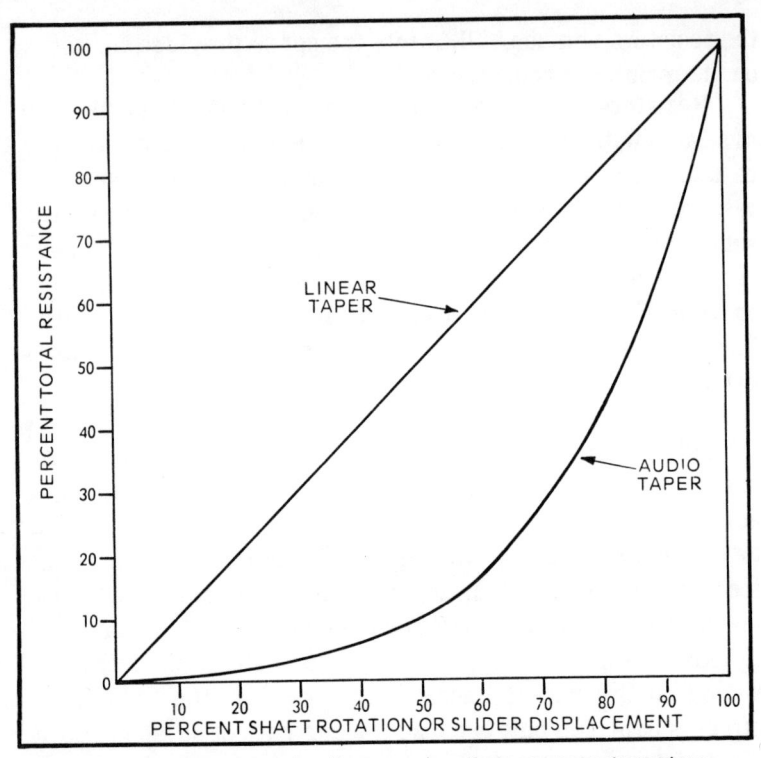

Fig. 5-3. A graph comparing linear and audio taper potentiometers.

capacitors. For units rated at 10 percent tolerance, the nominal values in Table 5-1 apply, except that appropriate dimensions are picofarads and microfarads. Incidentally, you may find the expression "micromicrofarads" and the abbreviation "mmf" used to refer to picofarads in books written before the IEEE standardized nomenclature in the early 1960s.

In some cases, you have wide latitude in selecting values of capacitance. In power supply filters, for example, anything larger than about 30 uF will work well.

As far as the dielectric material used to make the capacitor is concerned, paper, Mylar, ceramic, and mica capacitors of the same value are all interchangeable. Be careful, though, that there is enough room on a circuit board or inside an enclosure for a substitute capacitor. A paper dielectric capacitor may be several times bigger than a ceramic unit.

In most circuits, you may **not** substitute an electrolytic capacitor for another type. Electrolytics generally have too much leakage resistance to be useful in applications for which they were not designed. Another precaution is not to use electrolytics with a voltage well below their rated level. The operation of these capacitors involves using the applied voltage to ionize the electrolyte in them, and this requires the application of a certain minimum voltage.

Unlike resistors, capacitors are not rated in terms of their power dissipation (in watts). Instead, they are rated in terms of the maximum dc voltage that can be placed across them. This **dc working voltage** is abbreviated **WVDC**. Voltages above this level are likely to cause a breakdown of the dielectric and permanent damage to the capacitor.

You can connect several capacitors in series to obtain a voltage rating greater than you could get with a single unit, but you must remember that you decrease the total capacitance as you add capacitors in series. In addition, it is wise to use resistors of 50K or so in series with each capacitor to equalize the voltage drops.

Generally, it is not practical to change the value of a variable capacitor used for tuning by connecting fixed capacitors in series or parallel with it, because this also diminishes the range of the capacitor and, consequently, the frequency range that can be tuned. If you desire to reduce the capacitance of a variable capacitor, you **can** remove plates.

COILS AND TRANSFORMERS

When a coil is used in a tuned circuit, its inductance is, of course, critical in determining the resonant frequency of that circuit. In other applications, though, coils may be used to block ac while passing dc. In these **choke** coil applications, the exact value of inductance is not critical, and other chokes of greater inductance may be readily substituted.

A critical consideration in making these substitutions is the current-carrying ability of the coil. Never substitute a coil of lower current capacity than that specified, without checking first to insure that the current actually flowing in the circuit is within the rating of the new coil.

Power transformers are rated in terms of both **volt-amperes** (VA), which is the product of maximum volts times maximum amperes that can be supplied by the transformer, and in terms of watts, which is voltamperes times the cosine of the phase angle between the voltage and the current. For most electronic projects, loads are nonreactive, voltage and current are in phase, and voltamperes and watts are equal.

If you need for a ham transmitter a power transformer with a certain power rating, you may be able to substitute a TV-type transformer of considerably lower rating. These home entertainment units are rated for continuous duty, while transmitter duty is intermittent. As a rule of thumb, you can increase the power dissipation of home entertainment transformers 40 percent for service in amateur transmitters. Note that this means an increase in **current** capacity; the voltage is fixed by the ratio of turns in the primary and secondary windings.

Another important rating for power transformers is their design frequency. Conventional transformers used in television sets and common electrical devices are intended to work at 60 Hz. You will occasionally find transformers in surplus stores, however, that were designed for aircraft electrical systems that operate at 400 Hz. There is no way you can use these transformers at 60 Hz. Their reactance at this frequency is so low that they will draw excessive current and burn up. On the other hand, 60 Hz transformers work quite well at 400 Hz, but that isn't likely to do you much good.

As for the actual voltage rating of power transformers, it is best not to deviate too much from the values specified in a project parts list. If you just happen to have a transformer that has an output voltage only 10 percent or so from the specified level, you shouldn't have any trouble substituting it, but voltages that vary widely from the recommended value should be avoided.

Impedance-matching transformers are more critical than power transformers. There are wide gaps between impedance values, so it isn't possible to select the next higher or lower unit. In terms of frequency, it is again not possible to substitute units with vastly different design frequencies for each other. In fact, the only real substitutions you can make between impedance-matching transformers are between units with identical characteristics but made by different manufacturers. Even here, it is well to make sure that the new unit will fit into the available space before going ahead with the swap.

DIODES

Modern circuits use silicon diodes for all functions. Older power supply circuits used selenium rectifier diodes, but these are completely obsolete, and circuits will operate more efficiently with silicon diodes of the proper current rating. Other circuits used germanium diodes for small-signal applications, but here again, silicon diodes are the modern choice. You may find a project in an old magazine that calls for some other kind of diode than silicon, but you won't find that diode available in the catalogs any more. In general, silicon diodes are superior to all others in terms of heat resistance and current capacity.

The important ratings in selecting diodes and in deciding whether substitutes will work are **peak inverse voltage**, abbreviated **PIV**, and **current capacity**. Current capacity is self-explanatory. The PIV is the maximum voltage that can be applied in the reverse direction across a diode junction without causing breakdown and permanent damage. Note that this is a **peak** rating. If your power supply transformer provides an output that is a 200V **rms**, the **peak** voltage is 1.4 times 200V, or **280V**.

Older designs use a number of diodes in series to provide increased PIV ratings. These circuits require resistors and capacitors across each diode to equalize the voltage drop across the stack. Presently, it is cheaper to buy rectifier diodes with very high PIV ratings than to buy several diodes and associated resistors and capacitors. Consequently, diode stacks are now used only in very high-voltage power supplies. As semiconductor technology advances, we can expect that even here, diode stacks will be replaced by single high-PIV diodes.

Zener diodes are in a different class than conventional diodes from a design standpoint. However, from a substitution standpoint, the same two considerations—voltage and current—apply. With zeners, the voltage is **zener breakdown** voltage, but the only difference is that this reverse voltage does not cause permanent damage to the device. As before, the current rating is self-explanatory.

BIPOLAR TRANSISTORS

When we look at the thousands of types of transistors listed in the various manufacturers' handbooks, we might wonder how one could ever decide which of these are interchangeable. It turns out, though, that the situation isn't as bad as it first appears. Look at the specifications for specific transistors: Current gains vary from 100 to 1000 percent! Other ratings change substantially with current gain or temperature. Actually, the chance of finding two perfectly matched transistors (outside an integrated circuit) is about the same as finding two identical snowflakes.

The key to circuit design for bipolar transistors is to bias them so that the variations in their characteristics have little if any effect on circuit performance. This allows you, as an experimenter, some wide latitude in substituting types.

Some manufacturers have simplified your task considerably. Both Motorola and RCA, for example, make lines of transistors specifically for experimenters, and they publish lists cross referencing their transistors and units for which they may be substituted. The RCA line of replacement transistors is designated "SK," and the Motorola hobbyist transistors are all identified with the prefix "HEP." Substitution guides are available from radio parts stores and mail order houses.

If you do not want to use one of the devices listed in the substitution guides, you can make a pretty good judgment based on the published transistor parameters. The best way to find out the complete "pedigree" of a transistor is to obtain the specification sheet published by the manufacturer. You can get sheets for specific transistors by writing to the manufacturer's home office. The addresses of these offices are published in a catalog you can find at any good library, called the **Electronic Engineers Master** catalog, or **EEM**.

If you are in a hurry or are too lazy to scare up a complete specification sheet, you can get a pretty good notion of the transistor characteristics from the data listed in the manufacturer's transistor handbook.

What data are you interested in? First, you want to know whether the device is pnp or npn; then whether it is germanium or silicon. Germanium transistors generally drop about 0.7V between base and emitter at room temperature, while silicons drop only about 0.2V. This has an influence on the way the engineer designs his circuit and, generally, a germanium transistor cannot be used to replace a silicon device.

The next important consideration is power dissipation. Unless a circuit has been grossly overdesigned, you cannot expect to replace a 20W transistor with a 10W transistor. While you are checking to see that your proposed substitute does not have a lower power rating than the transistor it is to replace, take a look at the maximum voltage ratings specified. If they are lower than those listed for the transistor in the parts list, you will have to check the circuit to be sure that they will not be exceeded.

You may be confused by the abbreviations used in the specification sheets and handbooks. Capital letters indicate dc levels; V for voltage, I for current, etc. Breakdown voltage is indicated in at least two ways, depending on the manufacturer. Sometimes, **BV** is used, and at other times, (**BR**) is used as a subscript of the letter "V."

Other subscripts refer to the transistor elements between which the voltage is measured or through which the current flows, and to the condition of the remaining element. For example, V_{CBO} refers to the voltage between the transistor collector and base, with the emitter **open**. And V_{CBS} would be the same voltage measured with the emitter **shorted** to the base. Other letters used for the third character of the subscript are "R" and "X" or "V." The subscript "R" means that a resistor is connected between the unspecified terminal and one of the others, and X and V means that a certain voltage is applied between the unspecified terminal and one of the others. The values of resistance or voltage must be indicated

159

adjacent to the listing. If you have to choose between subscripts, remember that the **worst case** is always the one in which the unspecified terminal is left floating.

If everything matches so far, check the range of values for **current gain.** This is abbreviated h_{FE} or β (beta). Most circuit designs are unaffected by high values of beta, but beware of replacement transistors with lower minimum beta than the specified transistor. Of course, if you have a transistor tester that measures beta, you may be able to use even these transistors, if their actual beta is greater than the minimum for which the circuit was designed.

If you are dealing with a higher frequency rf application, you will also be interested in the **frequency** performance of your replacement transistor. The most commonly listed frequency parameter of transistors is the **gain-bandwidth product**, designated f_T. This is the frequency at which the voltage gain of the transistor in a common-emitter amplifier becomes unity. Obviously, you do not want a transistor with a value of f_T near the frequency at which you will be operating.

If you are substituting transistors in a switching circuit, an additional characteristic you will want to look at is the voltage drop from collector to emitter when the transistor is in the **on,** or **saturated** state. This is designated "$V_{CE(SAT)}$." It should be as low in your replacement, or lower, than it was in the original transistor.

FIELD-EFFECT TRANSISTORS

As with bipolar transistors, RCA and Motorola offer replacement FET types designated, respectively, "SK" and "HEP." The same substitution guides may be used for FETs as were used for bipolars.

If you are not using a substitution guide, the first thing to check is the **channel polarity** of the FET you want to replace. Once you have decided whether it is n-channel or p-channel, you will have to decide whether you have a junction FET or a depletion or enhancement MOSFET. The three types of FETs cannot be interchanged without changing other circuit elements to alter the bias. As with other components we have been discussing, it is important that you do not exceed the power rating or the maximum voltage and current ratings of your replacement FET.

The performance characteristics of FETs vary considerably among devices with the same part number, and circuits that employ FETs are designed with this in mind. This lets you substitute quite widely, within the constraints mentioned above.

While it is possible that some circuits will work with grossly dissimilar transistors (sometimes you can even substitute a dual-gate MOSFET with both gates tied together for a single-gate unit), for best results, you should consult the transistor specification sheet or handbook listings to learn the nominal values of the two most important FET parameters, I_{DSS} and $V_{GS(OFF)}$. The first of these is the current that flows from source to drain when the gate is shorted to the source and the indicated voltage is applied. And $V_{GS(OFF)}$ is the gate-to-source voltage required to pinch off the channel and prevent any current from flowing from drain to source. It is also equal to the minimum drain-to-source voltage that will cause a current equal to I_{DSS} to flow when the gate is shorted to the source. Sometimes, I_{DSS} and $V_{GS(OFF)}$ are not published in the handbook tables, but **drain characteristic curves are.**

Figure 5-4 shows how to find I_{DSS} and $V_{GS(OFF)}$ from the drain characteristic curves. To find I_{DSS} and the drain-to-source voltage equal to $V_{GS(OFF)}$ from the FET drain curves, first find the curve representing $V_{GS} = 0$. Then extend the straight portions of both regions of the curve until they cross. The drain current at this point is I_{DSS}. The drain-to-source voltage equals $-V_{GS(OFF)}$.

UNIJUNCTION TRANSISTORS

Compared to other types of transistors, there are not many types of unijunctions. Most UJT oscillator circuits are not at all critical, and UJTs can be substituted with relative ease.

INTEGRATED CIRCUITS

Different types of digital ICs that perform the same function can be substituted quite readily, especially by ex-

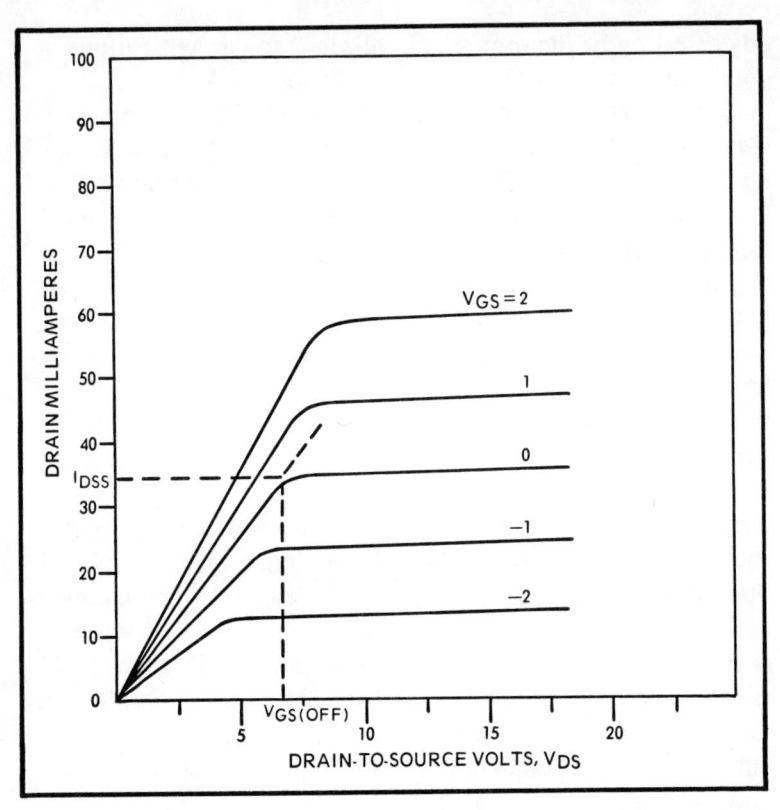

Fig. 5-4. A family of drain characteristic curves.

perimenters, who are not building computers with high pulse repetition rates. The most important consideration is to match ICs for supply voltages and logic 1 and 0 levels. Most of the digital logic ICs that you will encounter use **transistor-transistor logic**, abbreviated **TTL**. This type of device requires a 5V dc supply. Some logic ICs are of the older **resistor-transistor logic**, or **RTL** type. These devices require a 3.6V dc supply and are not readily matched to TTL devices.

Linear ICs such as operational amplifiers and differential amplifiers can be readily substituted among themselves, as long as care is taken to provide the appropriate feedback and frequency-compensating connections for each type of amplifier. For example, Fairchild 741 op amps can be readily substituted for that manufacturer's 709 type op amps—without the external frequency compensation required by the 709s.

162

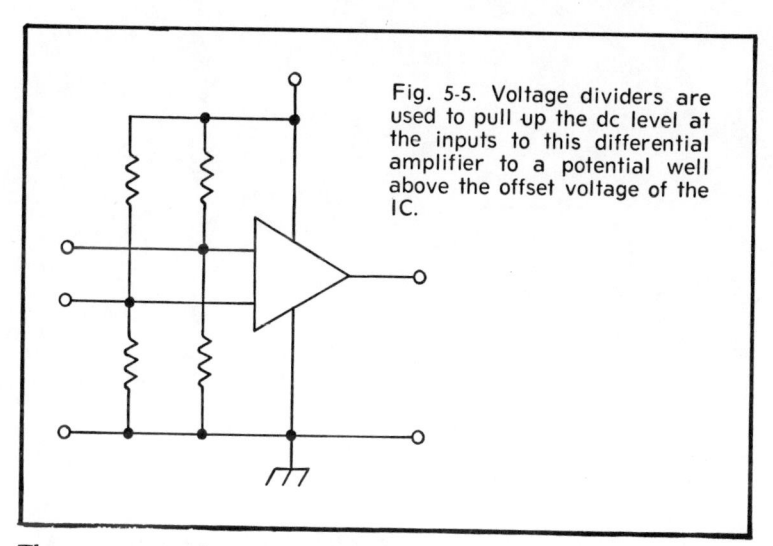

Fig. 5-5. Voltage dividers are used to pull up the dc level at the inputs to this differential amplifier to a potential well above the offset voltage of the IC.

The reverse is also true, as long as you **do** provide the external frequency-compensating network.

In substituting some differential amplifiers and operational amplifiers, it is possible to simply "float" the inputs; that is, to leave them isolated from ground. However, this can introduce distortion if the signal level is low and the amplifier has a large **offset** voltage. (The offset voltage is a dc voltage that appears in series with either input lead of the amplifier, and is an inherent characteristic of the IC.) It is usually better to use a voltage divider, as shown in Fig. 5-5, to hold the inputs above ground at some level greater than the offset voltage. This practice is called **pulling up** the input, and the voltage divider in this application is called a **pullup** circuit.

Appendixes

Appendix A:
Color Codes for Electronic Components & Wiring

Appendix B:
Electronic Symbols Used In Schematics

Electronic Color Coding APPENDIX A

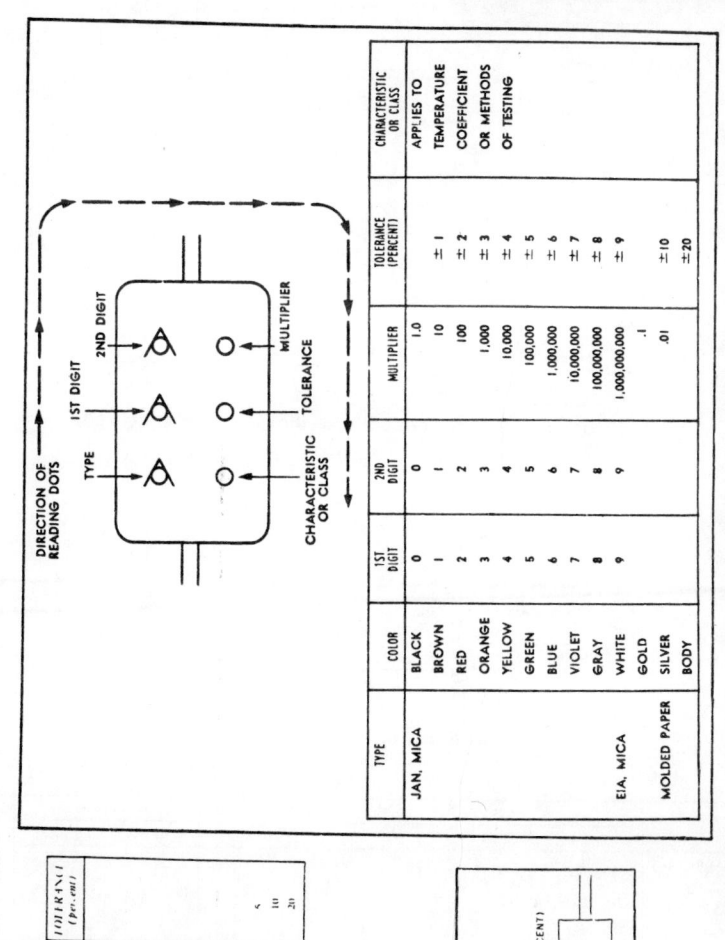

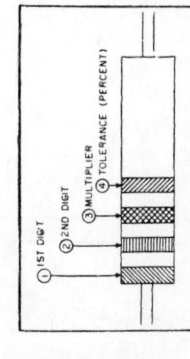

COLOR	1ST DIGIT	2ND DIGIT	MULTIPLIER	TOLERANCE (percent)
Black	0	0	1	
Brown	1	1	10	
Red	2	2	100	
Orange	3	3	1,000	
Yellow	4	4	10,000	
Green	5	5	100,000	
Blue	6	6	1,000,000	
Violet	7	7	10,000,000	
Gray	8	8	100,000,000	
White	9	9	1,000,000,000	
Gold			.1	5
Silver			.01	10
No color				20

A

B

Fig. A1. Resistor color code.

TYPE	COLOR	1ST DIGIT	2ND DIGIT	MULTIPLIER	TOLERANCE (PERCENT)	CHARACTERISTIC OR CLASS
JAN. MICA	BLACK	0	0	1.0		APPLIES TO TEMPERATURE COEFFICIENT OR METHODS OF TESTING
	BROWN	1	1	10	±1	
	RED	2	2	100	±2	
	ORANGE	3	3	1,000	±3	
	YELLOW	4	4	10,000	±4	
	GREEN	5	5	100,000	±5	
	BLUE	6	6	1,000,000	±6	
	VIOLET	7	7	10,000,000	±7	
EIA MICA	GRAY	8	8	100,000,000	±8	
	WHITE	9	9	1,000,000,000	±9	
	GOLD			.1	±10	
MOLDED PAPER	SILVER			.01	±20	
	BODY					

Fig. A2. 6-Dot color code for mica and molded paper capacitors.

167

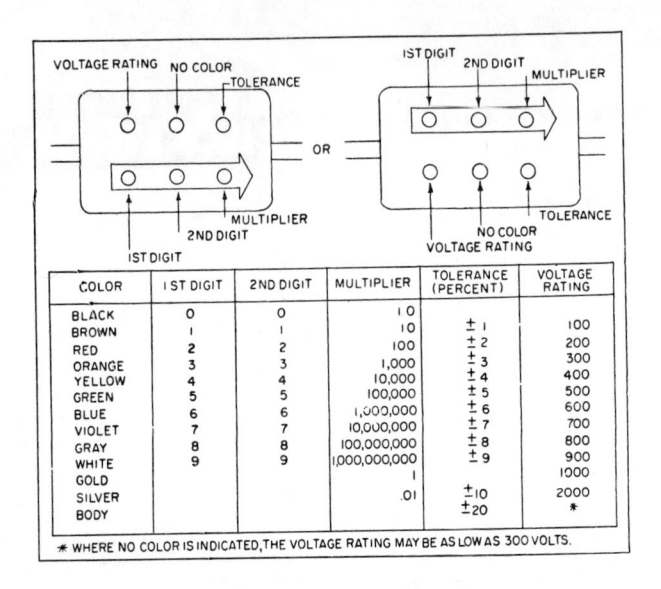

COLOR	1ST DIGIT	2ND DIGIT	MULTIPLIER	TOLERANCE (PERCENT)	VOLTAGE RATING
BLACK	0	0	1.0		
BROWN	1	1	10	±1	100
RED	2	2	100	±2	200
ORANGE	3	3	1,000	±3	300
YELLOW	4	4	10,000	±4	400
GREEN	5	5	100,000	±5	500
BLUE	6	6	1,000,000	±6	600
VIOLET	7	7	10,000,000	±7	700
GRAY	8	8	100,000,000	±8	800
WHITE	9	9	1,000,000,000	±9	900
GOLD			1		1000
SILVER			.01	±10	2000
BODY				±20	*

* WHERE NO COLOR IS INDICATED, THE VOLTAGE RATING MAY BE AS LOW AS 300 VOLTS.

Fig. A3. 5-Dot color code for capacitors (dielectric not specified).

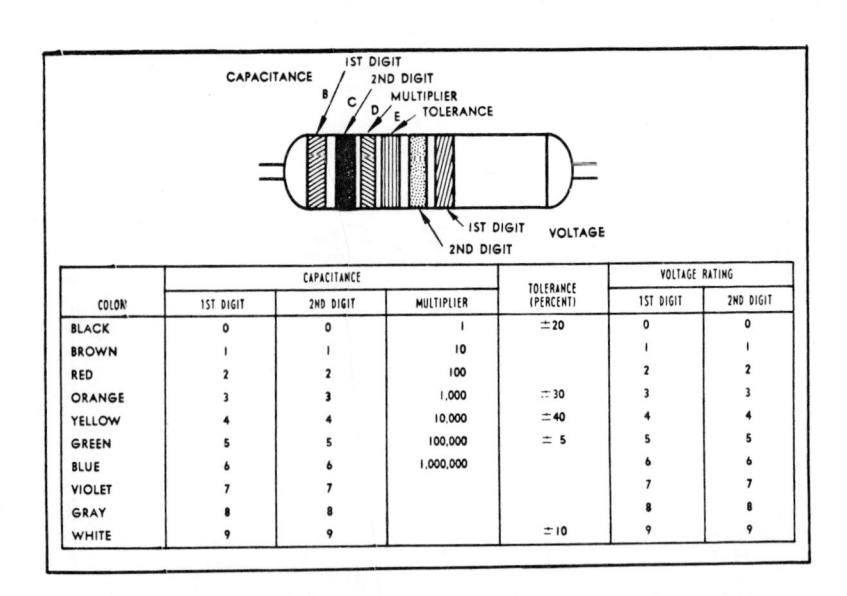

COLOR	CAPACITANCE 1ST DIGIT	CAPACITANCE 2ND DIGIT	MULTIPLIER	TOLERANCE (PERCENT)	VOLTAGE RATING 1ST DIGIT	VOLTAGE RATING 2ND DIGIT
BLACK	0	0	1	±20	0	0
BROWN	1	1	10		1	1
RED	2	2	100		2	2
ORANGE	3	3	1,000	±30	3	3
YELLOW	4	4	10,000	±40	4	4
GREEN	5	5	100,000	±5	5	5
BLUE	6	6	1,000,000		6	6
VIOLET	7	7			7	7
GRAY	8	8			8	8
WHITE	9	9		±10	9	9

Fig. A4. 6-Band color code for tubular paper dielectric capacitors.

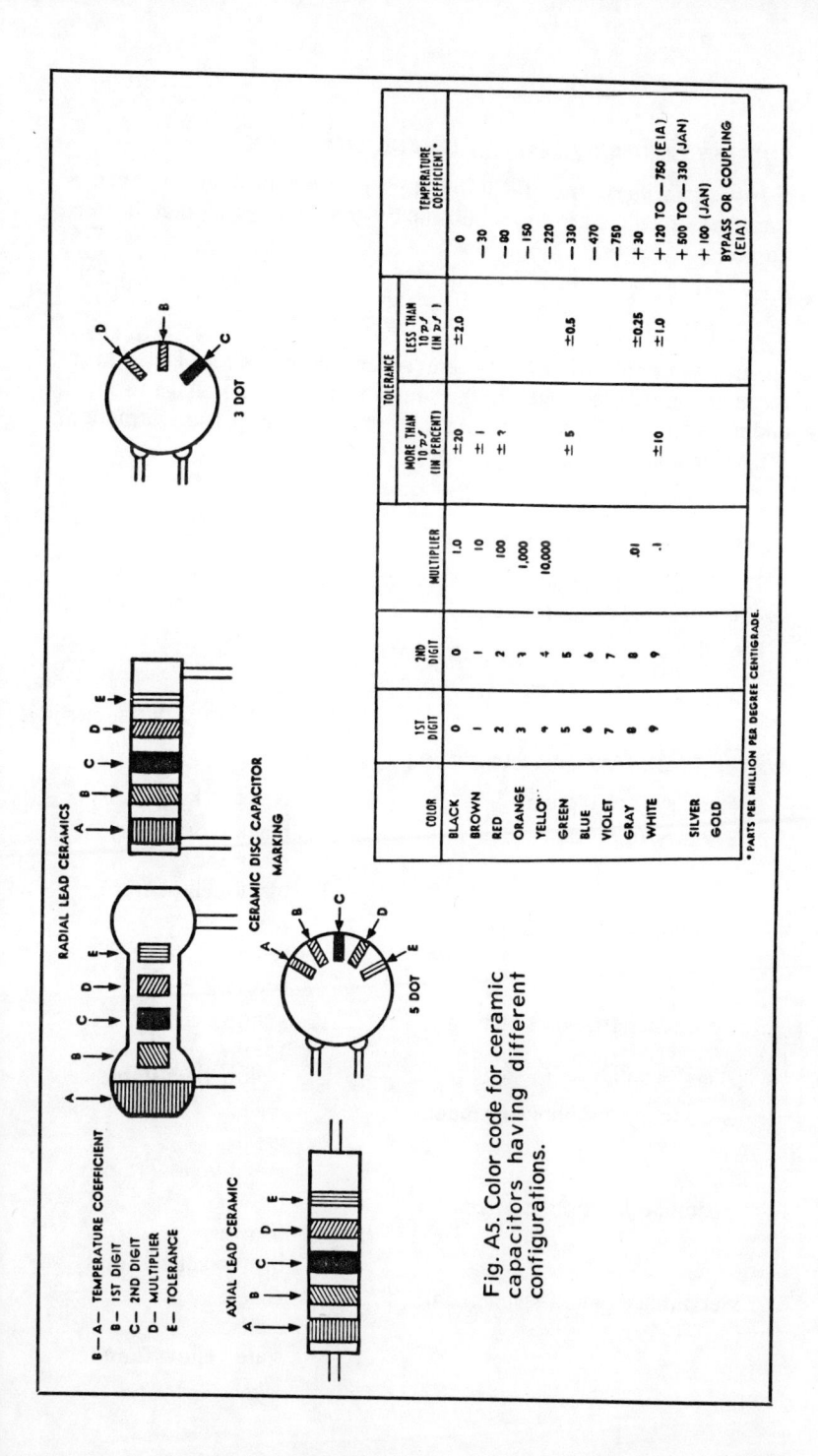

B—
- A— TEMPERATURE COEFFICIENT
- B— 1ST DIGIT
- C— 2ND DIGIT
- D— MULTIPLIER
- E— TOLERANCE

RADIAL LEAD CERAMICS

AXIAL LEAD CERAMIC

CERAMIC DISC CAPACITOR MARKING

3 DOT

5 DOT

COLOR	1ST DIGIT	2ND DIGIT	MULTIPLIER	TOLERANCE MORE THAN 10 pf (IN PERCENT)	TOLERANCE LESS THAN 10 pf (IN pf)	TEMPERATURE COEFFICIENT*
BLACK	0	0	1.0	±20	±2.0	0
BROWN	1	1	10	±1		−30
RED	2	2	100	±2		−80
ORANGE	3	3	1,000			−150
YELLOW	4	4	10,000			−220
GREEN	5	5		±5	±0.5	−330
BLUE	6	6				−470
VIOLET	7	7				−750
GRAY	8	8	.01		±0.25	+30
WHITE	9	9	.1	±10	±1.0	+120 TO −750 (EIA)
SILVER						+500 TO −330 (JAN)
GOLD						+100 (JAN)
						BYPASS OR COUPLING (EIA)

*PARTS PER MILLION PER DEGREE CENTIGRADE.

Fig. A5. Color code for ceramic capacitors having different configurations.

CURRENT TRANSFORMERS, STEP UP AND STEP DOWN

In transformers, the effects on current and voltage are inverse. A transformer will step down current by the same ratio that it steps up voltage.

$$\frac{Ip}{Is} = \frac{Ns}{Np}$$

Ip and *Is* represent primary and secondary currents. The current formula can be rearranged in the same manner as the voltage formula, and in as many different ways. As in the case of voltage transformation, Ns/Np is the turns ratio.

IMPEDANCE TRANSFORMERS

The ratio of secondary to primary impedance of a transformer varies as the square of the turns ratio.

$$\frac{Zs}{Zp} = \frac{Ns^2}{Np^2}$$

POWER TRANSFORMER COLOR CODE

Primary (not tapped)	two black leads
Primary (tapped)	black (common)
	black-red
	black-yellow (tap)
Secondary (high voltage)	red
	red
	red-yellow (tap)
Secondary (rectifier filament)	yellow
	yellow
	yellow-blue (tap)
Secondary (amplifier filament)	green
	green
	green-yellow (tap)
Secondary (amplifier filament)	brown
	brown
	brown-yellow (tap)
Secondary (amplifier filament)	slate
	slate
	slate-yellow (tap)

In power transformers (Fig. A6) a tapped lead always has two colors and yellow is always one of these colors.

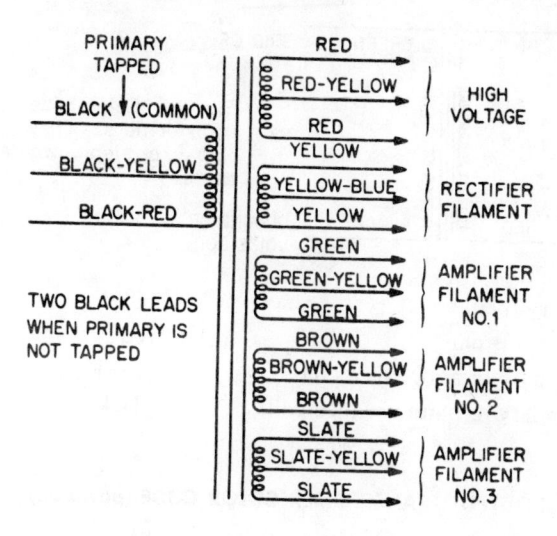

Fig. A6. Color code for power transformers.

I-F TRANSFORMER COLOR CODE

See Fig. A7.

Primary (plate)	blue
Primary (B-plus)	red
Secondary (grid or diode)	green
Secondary (grid or diode return, AVC, or ground)	black
Secondary (full-wave diode)	green-black (tap)

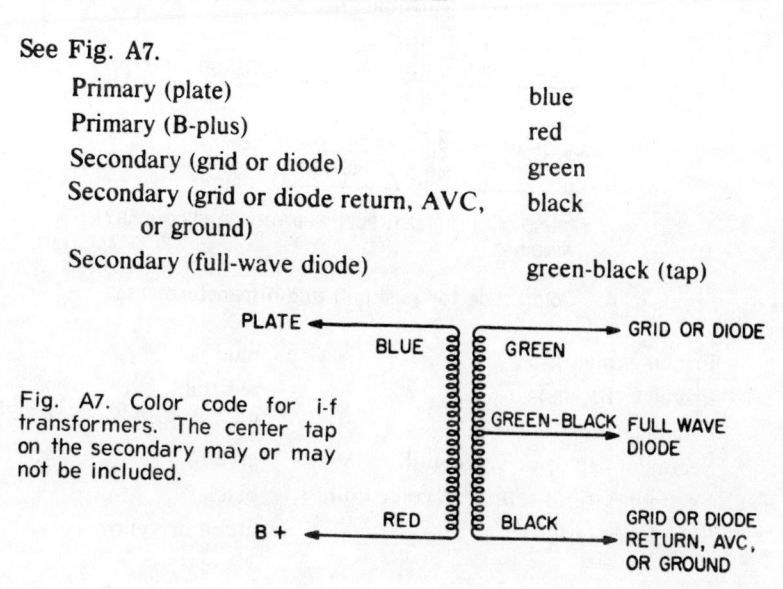

Fig. A7. Color code for i-f transformers. The center tap on the secondary may or may not be included.

AUDIO and OUTPUT TRANSFORMER COLOR CODE (single ended)

See Fig. A8.

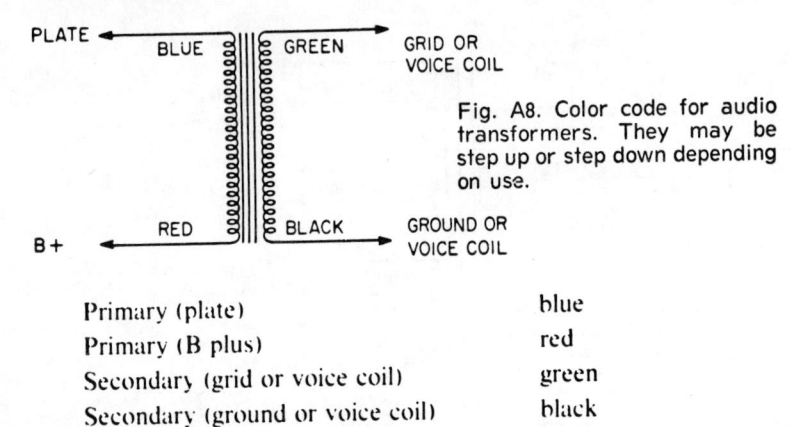

Fig. A8. Color code for audio transformers. They may be step up or step down depending on use.

Primary (plate)	blue
Primary (B plus)	red
Secondary (grid or voice coil)	green
Secondary (ground or voice coil)	black

AUDIO and OUTPUT TRANSFORMER COLOR CODE (pushpull)

See Fig. A9.

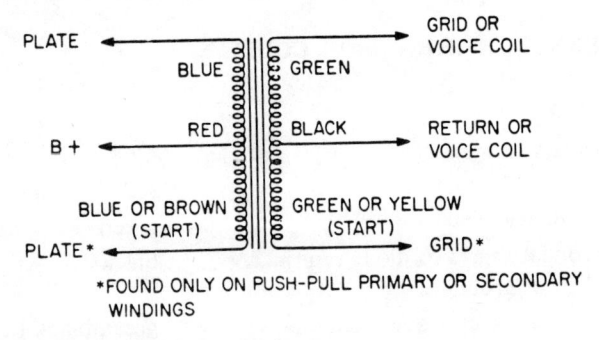

*FOUND ONLY ON PUSH-PULL PRIMARY OR SECONDARY WINDINGS

Fig. A9. Color code for pushpull audio transformers.

Primary (plate)	blue
Primary (B plus)	red (tap)
Primary (plate)	blue or brown
Secondary (grid or voice coil)	green
Secondary (grid return or voice coil)	black
Secondary (grid)	green or yellow

Electronic Symbols

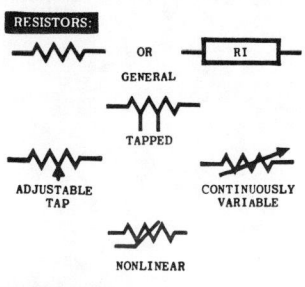

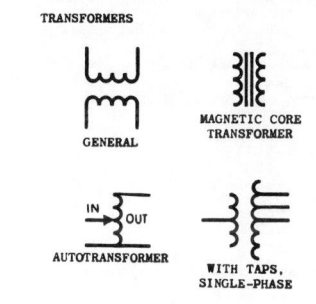

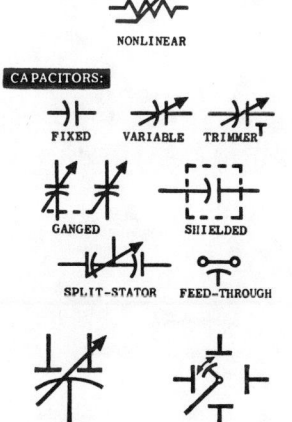

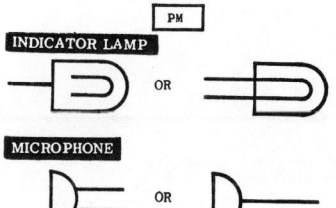

APPENDIX B

RESISTORS:
- GENERAL / OR R1
- TAPPED
- ADJUSTABLE TAP
- CONTINUOUSLY VARIABLE
- NONLINEAR

CAPACITORS:
- FIXED
- VARIABLE
- TRIMMER
- GANGED
- SHIELDED
- SPLIT-STATOR
- FEED-THROUGH
- DIFFERENTIAL
- PHASE SHIFT

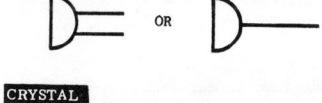

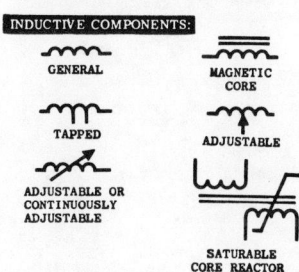

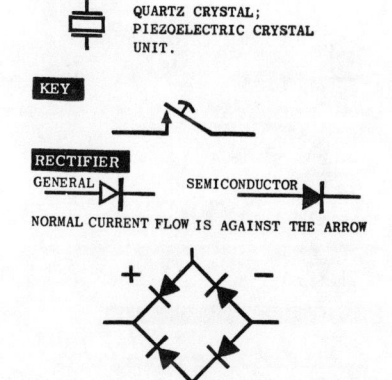

(WHEN CAPACITOR ELECTRODE IDENTIFICATION IS NECESSARY, THE CURVED ELEMENT SHALL REPRESENT THE OUTSIDE ELECTRODE IN FIXED PAPER-DIELECTRIC AND CERAMIC-DIELECTRIC, THE NEGATIVE ELECTRODE IN ELECTROLYTIC CAPACITORS THE MOVING ELEMENT IN VARIABLE AND ADJUSTABLE CAPACITORS, AND THE LOW POTENTIAL ELEMENT IN FEED-THROUGH CAPACITORS.)

INDUCTIVE COMPONENTS:
- GENERAL
- MAGNETIC CORE
- TAPPED
- ADJUSTABLE
- ADJUSTABLE OR CONTINUOUSLY ADJUSTABLE
- SATURABLE CORE REACTOR

TRANSFORMERS
- GENERAL
- MAGNETIC CORE TRANSFORMER
- AUTOTRANSFORMER (IN / OUT)
- WITH TAPS, SINGLE-PHASE

PERMANENT MAGNET
PM

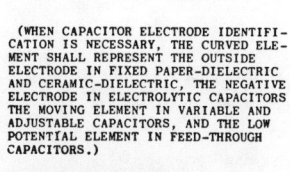

INDICATOR LAMP
OR

MICROPHONE
OR

CRYSTAL
QUARTZ CRYSTAL; PIEZOELECTRIC CRYSTAL UNIT.

KEY

RECTIFIER
GENERAL   SEMICONDUCTOR

NORMAL CURRENT FLOW IS AGAINST THE ARROW

FULL WAVE BRIDGE TYPE

AMPLIFIER

TRIANGLE POINTS IN DIRECTION OF TRANSMISSION (SIGNAL FLOW)

AMPLIFIER WITH EXTERNAL FEEDBACK PATH

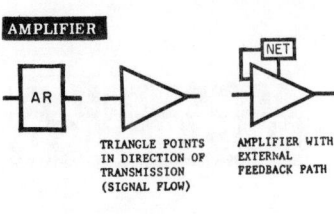

BASIC SYMBOL INDICATES ANY METHOD OF AMPLIFICATION EXCEPT THAT OPERATING ON THE PRINCIPLE OF ROTATING MACHINERY.

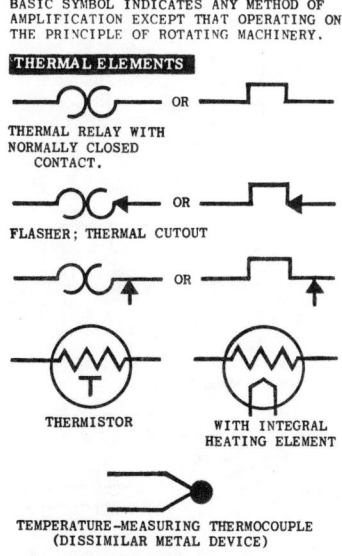

THERMAL ELEMENTS

OR

THERMAL RELAY WITH NORMALLY CLOSED CONTACT.

OR

FLASHER; THERMAL CUTOUT

OR

THERMISTOR

WITH INTEGRAL HEATING ELEMENT

TEMPERATURE-MEASURING THERMOCOUPLE (DISSIMILAR METAL DEVICE)

INPUTS (NONSTANDARD)

OR

PATH, TRANSMISSION

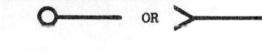

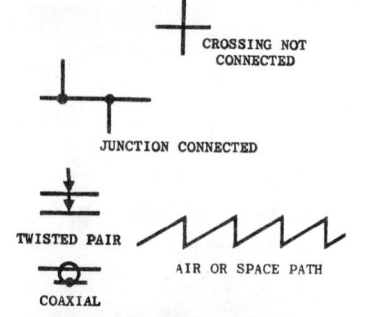

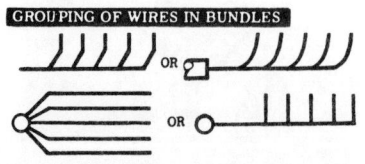

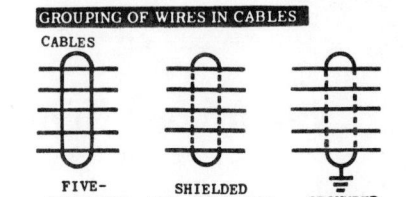

CROSSING NOT CONNECTED

JUNCTION CONNECTED

TWISTED PAIR

AIR OR SPACE PATH

COAXIAL

GROUPING OF WIRES IN BUNDLES

OR

OR

GROUPING OF WIRES IN CABLES

CABLES

FIVE-CONDUCTOR CABLE

SHIELDED FIVE-CONDUCTOR CABLE

GROUNDED SHIELD

NUMBER OF CONDUCTORS MAY BE ONE OR MORE AS NECESSARY

SWITCHES

GENERAL (SINGLE THROW)

GENERAL (DOUBLE THROW)

TWO POLE DOUBLE THROW SWITCH

KNIFE SWITCH

PUSHBUTTON (MAKE)

PUSHBUTTON (BREAK)

PUSHBUTTON TWO CIRCUIT

SELECTOR SWITCHES

OR

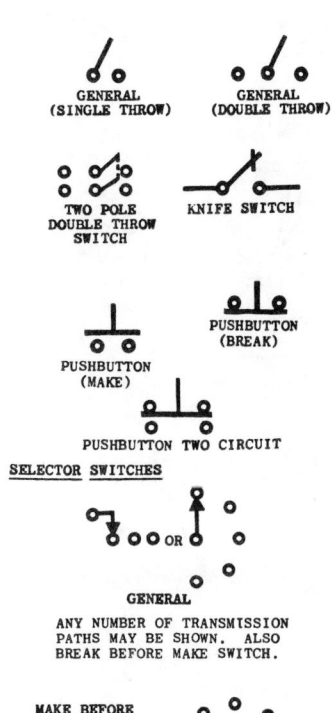

GENERAL

ANY NUMBER OF TRANSMISSION PATHS MAY BE SHOWN. ALSO BREAK BEFORE MAKE SWITCH.

MAKE BEFORE BREAK

WAFER, TYPICAL 3-POLE, 3-CIRCUIT SWITCH. VIEWED FROM END OPPOSITE CONTROL KNOB. FOR MORE THAN ONE SECTION, #1 IS NEAREST CONTROL KNOB.

CIRCUIT RETURNS

CHASSIS CONNECTION

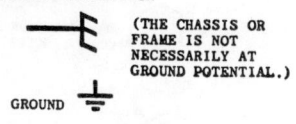

(THE CHASSIS OR FRAME IS NOT NECESSARILY AT GROUND POTENTIAL.)

GROUND

CONTACTS (ELECTRICAL)

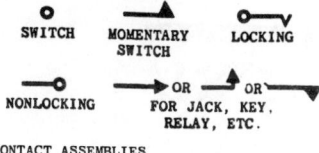

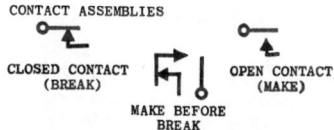

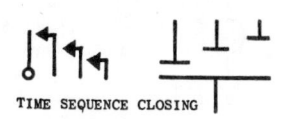

SWITCH   MOMENTARY SWITCH   LOCKING

NONLOCKING   OR   OR   FOR JACK, KEY, RELAY, ETC.

CONTACT ASSEMBLIES

CLOSED CONTACT (BREAK)   MAKE BEFORE BREAK   OPEN CONTACT (MAKE)

TIME SEQUENCE CLOSING

DISCONNECTING DEVICES

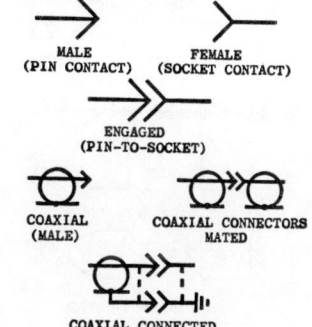

MALE (PIN CONTACT)   FEMALE (SOCKET CONTACT)

ENGAGED (PIN-TO-SOCKET)

COAXIAL (MALE)   COAXIAL CONNECTORS MATED

COAXIAL CONNECTED TO SINGLE CONDUCTOR

THE CONNECTOR SYMBOL IS NOT AN ARROWHEAD. IT IS LARGER AND THE LINES ARE DRAWN AT A 90° ANGLE.

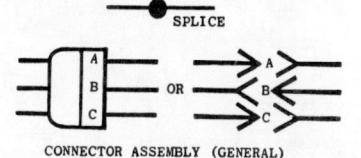

SPLICE

CONNECTOR ASSEMBLY (GENERAL)

ELECTRON TUBES

COMPONENT TUBE SYMBOLS

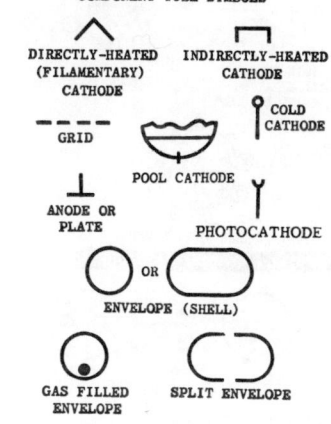

DIRECTLY-HEATED (FILAMENTARY) CATHODE   INDIRECTLY-HEATED CATHODE

GRID   POOL CATHODE   COLD CATHODE

ANODE OR PLATE   PHOTOCATHODE

ENVELOPE (SHELL)

GAS FILLED ENVELOPE   SPLIT ENVELOPE

SEMICONDUCTOR DEVICES

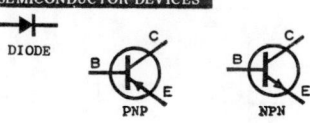

DIODE

PNP   NPN

TRANSISTORS

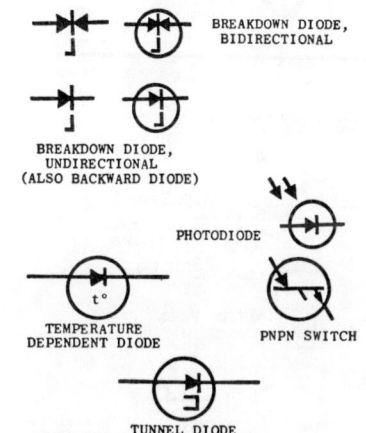

BREAKDOWN DIODE, BIDIRECTIONAL

BREAKDOWN DIODE, UNDIRECTIONAL (ALSO BACKWARD DIODE)

PHOTODIODE

TEMPERATURE DEPENDENT DIODE   PNPN SWITCH

TUNNEL DIODE

TYPICAL ELECTRON TUBES

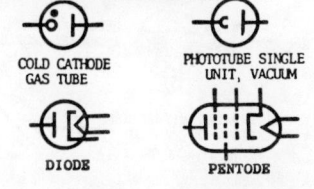

COLD CATHODE GAS TUBE   PHOTOTUBE SINGLE UNIT, VACUUM

DIODE   PENTODE

TYPICAL ELECTRON TUBES (Continued)

- TWIN TRIODE ILLUSTRATING ELONGATED ENVELOPE
- DIODE SHOWING BASE CONNECTIONS
- TWIN TRIODE WITH TAPPED HEATER

TYPICAL CATHODE RAY TUBES

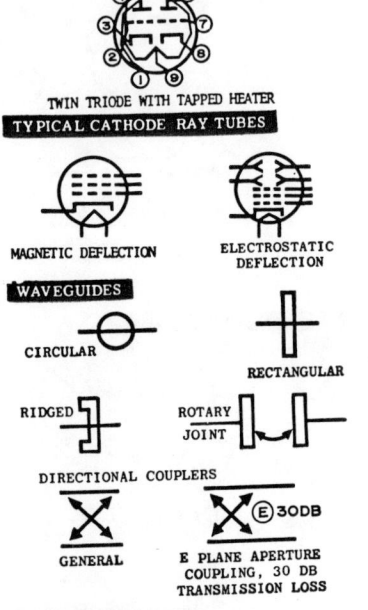

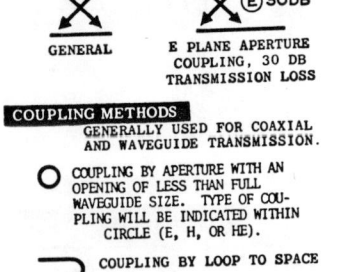

- MAGNETIC DEFLECTION
- ELECTROSTATIC DEFLECTION

WAVEGUIDES

- CIRCULAR
- RECTANGULAR
- RIDGED
- ROTARY JOINT

DIRECTIONAL COUPLERS

- GENERAL
- E PLANE APERTURE COUPLING, 30 DB TRANSMISSION LOSS

COUPLING METHODS

GENERALLY USED FOR COAXIAL AND WAVEGUIDE TRANSMISSION.

- O COUPLING BY APERTURE WITH AN OPENING OF LESS THAN FULL WAVEGUIDE SIZE. TYPE OF COUPLING WILL BE INDICATED WITHIN CIRCLE (E, H, OR HE).
- COUPLING BY LOOP TO SPACE
- COUPLING BY LOOP TO GUIDED TRANSMISSION PATH
- COUPLING BY PROBE FROM COAXIAL TO RECTANGULAR WAVEGUIDE WITH DIRECT-CURRENT GROUNDS CONNECTED

TYPICAL MAGNETRONS AND KLYSTRONS

- REFLEX KLYSTRON, APERATURE COUPLED
- TUNABLE MAGNETRON, APERTURE COUPLED

TYPICAL MAGNETRONS AND KLYSTRONS (Continued)

- RESONANT TYPE WITH COAXIAL OUTPUT
- TRANSMIT-RECEIVE (TR) TUBE GAS FILLED, TUNABLE INTEGRAL CAVITY, APERTURE COUPLED, WITH STARTER

ROTATING MACHINES

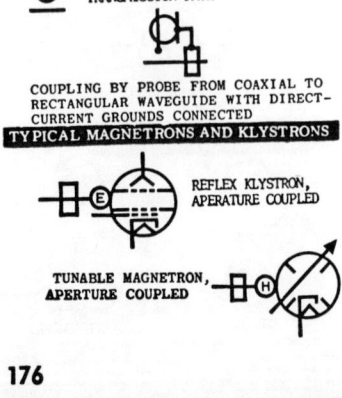

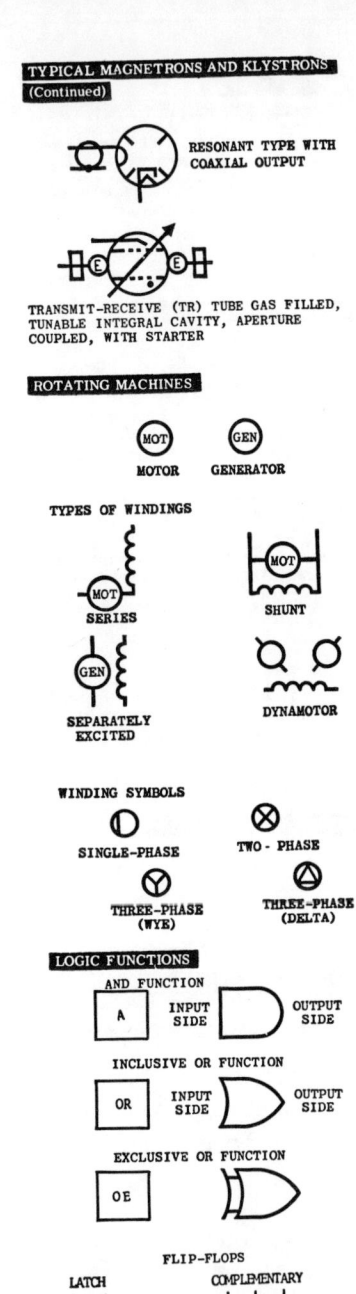

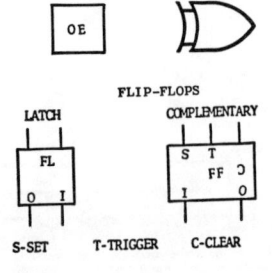

- MOTOR
- GENERATOR

TYPES OF WINDINGS

- SERIES
- SHUNT
- SEPARATELY EXCITED
- DYNAMOTOR

WINDING SYMBOLS

- SINGLE-PHASE
- TWO-PHASE
- THREE-PHASE (WYE)
- THREE-PHASE (DELTA)

LOGIC FUNCTIONS

AND FUNCTION
A — INPUT SIDE / OUTPUT SIDE

INCLUSIVE OR FUNCTION
OR — INPUT SIDE / OUTPUT SIDE

EXCLUSIVE OR FUNCTION
OE

FLIP-FLOPS

- LATCH — FL — 0, I
- COMPLEMENTARY — S T FF 0 — I

S-SET   T-TRIGGER   C-CLEAR

LOGIC FUNCTIONS (Continued)

NEGATION

ELECTRIC INVERTER

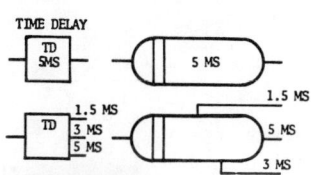

TIME DELAY

PICKUP HEADS

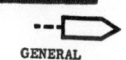

GENERAL

WRITING; RECORDING; HEAD, SOUND RECORDER

READING; PLAYBACK; HEAD, SOUND REPRODUCER

APPLICATION: WRITING, READING, AND ERASING

ERASING; ERASER, MAGNETIC

BATTERIES

ONE CELL   MULTICELL   TAPPED MULTICELL

(LONG LINE IS ALWAYS POSITIVE)

CIRCUIT PROTECTORS

FUSE

CIRCUIT PROTECTORS (Continued)

CIRCUIT BREAKERS

SWITCH

PUSH PULL OR PUSH

GANGED

ATTENUATORS

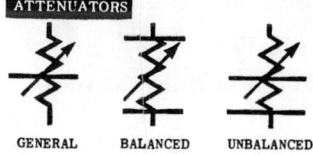

GENERAL   BALANCED   UNBALANCED

ANTENNAS

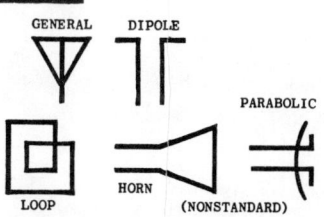

GENERAL   DIPOLE

LOOP   HORN   PARABOLIC (NONSTANDARD)

METERS

A — AMMETER
CRO — OSCILLOSCOPE
G — GALVANOMETER
MA — MILLIAMMETER
OHM — OHMMETER
V — VOLTMETER

HEADSET

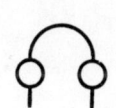

OR

INDEX

—A—

Acid
—bath 32
—core solder 17
—resistance material 32
Adhesive
—pads 30
—perforated board 29
Adjustable wrench 11
Air-wound coil 98
Alligator clip 19
Amberlith 36
Amplifier
—audio 21
—FET 151
Analyzer, specturm 117
Angle, conduction 68
Audio amplifier 21

—B—

Battery holders 107
Bench power supply 107
Bipolar transistor 158
BI-RBO 94
Bits 12
BJT-FET 108
—transistor-checker 128
BJT tests 129
Black
—Bakelite boxes 75
—plastic 76
Blank ripple 93
Bleeder resistor 112
Board
—copper-clad 35, 40
—mother 42
—perforated 27
—sensitized 35
Box, sloping-panel 73
Boxes
—black Bakelite 75
—plastic 74
—transparent plastic 75
—utility 73
Brackets, metal 41

Breadboards 27
Bulb, neon 52
Burnishing tools 32

—C—

Calibration 144
Capaci-bridge
—circuit 125
—construction 127
—operation 128
—project 121
—theory 123
Capacitors 153
—electrolytic 58
—troubleshooting 119
Capillarity 21
Cases 70
Cathode-ray tube 109
Chassis, metal 43
Checker, transistor 108
Chemical
—developing 38
—light-sensitive 35
Circuit
—capaci-bridge 125
—element 32
—firing 52
—integrated 101
—invertor 59
—parameter 82
—printed 32
—prototype 28
—randomizer 88
—speed controlled 67
Clip
—alligator 19
—paper 19
Coil 155
—air-wound 98
—trigger 52
—volt-amperes 156
Component 32
Conduction angle 68
Conductor 34
Connector, edge card 42

178

Construction
—capaci-bridge 127
—signal generator 141
—transistor-checker 135
Controller, triac 65
Control
—panel 76
—primary 76
Copper-clad board 35, 40
Copper
—ribber 29
—wire 46, 142
"Cordwood" packaging 31
Counter, ring 88

— D —

Dc power 111
"Debugging" 104
Decode-drivers 87, 115
Depletion MOSFET 132
Developing chemical 38
Diagonal side-cutting
  pliers 10
Dial, vernier 103
Digital display 106
Dikes 10
Diodes 157
—current capacity 157
—zener 157
—zener breakdown 158
Dip meter 108
Display
—digital 106
—LED 87
Doubler, voltage 41
Dress, lead 25
Drill, electric 12
Dry transfer 78

— E —

Edge
—card connector 42
—mounting 42
Electric drill 12
Electrolytic capacitor 58
Electronic voltmeter 105
Element, circuit 32
Equipment
—test 72
—troubleshooting 105
Etching 40
Eutectic solder 17
EVMs 105
Exposing 37

— F —

Faceplate 74
Feedbacks 76

FET
—amplifier 151
—preamplifier 96
—tests 133
FET-Set 95, 99
—operation 101
FETVMs 105
Field-effect transistor 105, 131, 151, 160
Files 14
—flat 15
—half-round 15
—rat 15
—round 15
Firing circuit 52
Flash
—bulb, xenon 52
—tube 49
Flat
—headed screwdrivers 11
—files 15
Fluorescent lamps 37
Frequencies, microwave 33
Frequency, intermediate 116
Flux, rosin 17

— G —

Gain, current 160
Generator, signal 106, 112, 139
Grid-dip meter 108
Gun 15
—soldering 15

— H —

Hacksaw 13, 45
Half-round files 15
Heatsink 19
High-power unit 152
Holders, battery 107
Hysteresis 69

— I —

Ic decoder-drivers 86
IEEE 154
Image, negative 37
Integrated circuits 161
Intermediate frequency 116
Inductors, troubleshooting 119
Inverter
—circuit 59
—project 56
Iron 15
—soldering 9

— J —

Joint
—rosin 20
—solder 17

179

—K—

Keyhole saw	13
Knives	14
Knob	
—turn-counting	81
—vernier	81
Knurled shaft	44

—L—

Labels, legible	77
Laboratory	36
Lamps	
—fluorescnet	37
—reflector	38
—test	93, 113
Lead dress	25
LED display	87
Legible labels	77
Light-sensitive chemical	35
Long nose pliers	9
Low-power unit	152

—M—

Machine, wave-soldering	41
Material, acid-resistant	32
Mechanical negative	38
Metal	
—brackets	41
—chassis	43
Meters	105
—dip	108
—grid-dip	108
—scales	76, 83
Micromicrofarads	154
Microwave frequencies	33
Modulator	116
MOSFET	26, 95, 132
—depletion	132
Mother board	42
Mounting, edge card	42
Multi-Strobe	48, 52
—construction	53
—operation	54
Miniboxes	71

—N—

Negative	
—image	37
—mechanical	36
—photo	35
Neon bulb	52
Neutralizing	115
Nibbler	13
—tool	13
Nonshorting type	82
Numerical readouts	83

—O—

Opaque bottom	75
Operation	
—capaci-bridge	128
—FET-Set	101
—signal generator	148
—transistor-checker	136
Oscilloscope	109

—P—

Packaging, "cordwood"	31
Pads, adhesive	30
Paper clip	19
Parameter, circuit	82
Pencil, soldering	16
Perforated board	27
Phillips head screwdriver	11
Photo negative	35
Picofarads	154
Plastic	
—black	76
—boxes	74
Pliers	9
—diagonal side cutting	10
—vice grip	11
—long nose	9
Point size	79
Polarity	111
Potentiometer	13, 44
Power	
—dc	111
—loss, resistors	152
—supply	49
Preamplifier, FET	96
Primary controls	76
Printed circuit	32
—master	30
Project	
—capaci-bridge	121
—inverter	56
—randomizer	86
—signal generator	137
Prototype circuit	28
Push-in terminals	28

—R—

Randomizer	
—circuit	88
—project	86
Rat files	15
Razor, X-acto keyhole	75
RBI	93
Readouts, numerical	83
Reflector lamp	38
Resistor	
—bleeder	112, 151
—power loss	152

—troubleshooting	119
Rf signals	116
Ribbon, copper	29
Ring counter	88
Ripple blank	93
Ripple-blanking input	93
Rosin	
—flux	17
—joint	20
Round files	15
Rubylith	36

—S—

Safelight	37
Sansitized board	35
Sans serif type	79
Saws	13
—keyhole	13
SCR	115
Scales, meter	83
Screwdrivers	
—flat headed	11
—Phillips head	11
Semiconductors, troubleshooting	119
Shaft	
—knurled	44
—variable-capacitor	45
Shears, tool-steel	13
Shorting type	82
Short wave receiver, FET-Set	95
Sidebands	116
Signal generator	106, 112, 139
—construction	141
—project	137
—operation	148
Slide switch	82
Signal	
—rf	116
—tracers	107
Slipups	104
Sloping panel box	73
Socket	
—punches	9
—transistor	26
Solder	110
—acid core	17
—eutectic	17
—joints	17
Soldering	
—irons	9
—gun	15
—pencil	16
Solid wire	46
Spectrum analyzer	117
Speed controller circuit	67
Spring clips	43
Strippers, wire	10
Supply	
—bench power	107
—power	49

Surface tension	21
Switches, toggle	13, 77
Systematic troubleshooting	110

—T—

Tantalum	57
Taper	153
Tension, surface	21
Terminals	
—push-in	28
—Vcc	113
Test	
—BJT	129
—equipment	72
—FET	133
—lamp	93, 113
Tickler	96
Tin snips	9
Toggle switches	13, 77
Tool	
—burnishing	32
—nibbling	13
Tool-steel shears	13
Tracers, signal	107
Transformers	155
Transistor	
—bipolar	158
—checker	108
—field-effect	131, 151, 160
—socket	26
—unijunction	161
Transistor-checker	
—BJT-FET	128
—construction	135
—operation	136
Transparent plastic boxes	75
Triac controller	65
Trigger coil	52
Tri-X	48
Troubleshooting	104
—capacitors	119
—equipment	105
—inductors	119
—resistors	119
—semiconductors	119
—systematic	110
—vacuum tubes	120
Tube	
—cathode-ray	109
—flash	49
Turn-counting knob	81
Type	
—nonshorting	82
—sans serif	79
—shorting	82

—U—

UHF	33
Unijunction transistor	161

Unit
—high-power 152
—low-power 152
Utility boxes 73

—V—

Vacuum tubes 42
—troubleshooting 120
—voltmeter 105
Variable-capacitor shafts 45
Vcc terminals 113
Vernier
—dial 103
—knobs 81
Vice-grip pliers 11
Voltage doubler 51
Volt-amperes, coils 156
Voltmeter 105
—electronic 105
—vacuum tube 105
Volt-ohm-milliameter 105

VOMs 105, 106
VTVMs 105

—W—

Wavemeter 108
Wave-soldering machine 41
Wire
—copper 142
—solid 46
—strippers 10
Wrench, adjustable 11

—X—

X-acto keyhole razor 75
Xenon flash bulb 52

—Z—

Zener break-down diodes 157, 158